T0310176

Beam Structures

Beam Structures

Classical and Advanced Theories

Erasmo Carrera
Politecnico di Torino, Italy

Gaetano Giunta
Centre de Recherche Public Henri Tudor, Luxembourg

Marco Petrolo
Politecnico di Torino, Italy

A John Wiley & Sons, Ltd., Publication

This edition first published 2011

Registered office
John Wiley & Sons Ltd, The Atrium, Southern Gate, Chichester, West Sussex, PO19 8SQ, United Kingdom

For details of our global editorial offices, for customer services and for information about how to apply for permission to reuse the copyright material in this book please see our website at www.wiley.com.

Library of Congress Cataloguing-in-Publication Data

Carrera, Erasmo.
Beam structures : classical and advanced theories / Erasmo Carrera, Gaetano Giunta,
Marco Petrolo. – 1st ed.
p. cm.
Includes bibliographical references and index.
ISBN 978-0-470-97200-7 (hardback)
1. Girders. I. Giunta, Gaetano. II. Petrolo, Marco. III. Title.
TA492.G5C37 2011
624.1′7723–dc23

2011019533

A catalogue record for this book is available from the British Library.

Print ISBN:9780470972007
ePDF ISBN: 9781119978572
Obook ISBN: 9781119978565
ePub ISBN: 9781119951049
Mobi ISBN: 9781119951056

Typeset in 10/12pt Times by Aptara Inc., New Delhi, India
Printed and bound in Singapore by Markono Print Media Pte Ltd

Contents

About the Authors

Erasmo Carrera

After earning two degrees (Aeronautics, 1986, and Aerospace Engineering, 1988) at the Politecnico di Torino, Erasmo Carrera received his PhD degree in Aerospace Engineering jointly at the Politecnico di Milano, Politecnico di Torino, and Università di Pisa in 1991. He began working as a Researcher in the Department of Aerospace for the Politecnico di Torino in 1992 where he held courses on Missiles and Aerospace Structure Design, Plates and Shells, and the Finite Element Method. He became Associate Professor of Aerospace Structures and Computational Aeroelasticity in 2000, and Full Professor at the Politecnico di Torino in 2011. He has visited the Institute für Statik und Dynamik, Universität Stuttgart twice, the first time as a PhD student (six months in 1991) and then as Visiting Scientist under a GKKS Grant (18 months in 1995–1996). In the summers of 1996, 2003 and 2009, he was Visiting Professor at the ESM Department of Virginia Tech, at SUPMECA in Paris (France) and at the CRP TUDOR in Luxembourg, respectively. His main research topics are: inflatable structures, composite materials, finite elements, plates and shells, postbuckling and stability, smart structures, thermal stress, aeroelasticity, multibody dynamics, and the design and analysis of non-classical lifting systems. He is author of more than 300 articles on these topics, many of which have been published in international journals. He serves as referee for international journals, and as a contributing editor for Mechanics of Advanced Materials and Structures, Composite Structures and Journal of Thermal Stress.

Gaetano Giunta

Gaetano Giunta graduated in Aerospace Engineering at the Politecnico di Torino in 2004. In 2007 he defended his PhD thesis on "Deterministic and Stochastic Hierarchical Analysis of Failure and Vibration of Composites Plates and Shells" at the Politecnico di Torino. Dr Giunta carried out his post-doc at the Centre de Recherche Public Henri Tudor in Luxembourg and the Politecnico di Torino from February 2008 to January 2010. Currently he is an R&D Engineer at the

Centre de Recherche Public Henri Tudor and is working on the project FNR CORE C09/MS/05 FUNCTIONALLY on "Functionally Graded Materials: Multi-Scale Modelling, Design and Optimisation" funded by the Fonds National de la Recherche Luxembourg (FNR). Dr Giunta would like to acknowledge the FNR for its support. His research covers the formulation of hierarchical analytical and finite element models for the static, free vibration, buckling and failure analysis of beam, plate, and shell structures made of conventional and advanced materials.

Marco Petrolo
Marco Petrolo is an Aerospace Engineer and a PhD student at the Politecnico di Torino. He has a BSc in Aerospace Engineering at the Politecnico di Torino and an MSc degree in Aerospace Engineering in a joint program between the Politecnico di Torino, TU Delft, and EADS. He is a Fulbright alumnus, and, as a such, he has spent research periods at the San Diego State University and at the University of Michigan. His research activity is focused on the development of refined models for the structural and aeroelastic design of composite and metallic structures. He works in Professor Carrera's research group in Turin on different aerospace applications, such as the structural analysis of composite lifting surfaces, multiscale problems, and nonlinear problems.

Preface

Beam models have made it possible to solve a large number of engineering problems over the last two centuries. Early developments, based on kinematic intuitions (bending theories), by pioneers such as Leonardo da Vinci, Euler, Bernoulli, Navier, and Barre de Saint Venant, have permitted us to consider the most general three-dimensional (3D) problem as a one-dimensional (1D) problem in which the unknowns only depend on the beam-axis position. These early theories are known as engineering beam theories (EBTs) or the Euler–Bernoulli beam theory (EBBT). Recent historical reviews have proposed that these theories should be referred to as the DaVinci–Euler–Bernoulli beam theory (DEBBT). The drawbacks of EBT are due to the intrinsic decoupling of bending and torsion (cross-section warping is not addressed by EBT) as well as to the difficulties involved in evaluating the additional five (normal and shear) stress components that are not provided by the Navier formula. Many torsion-beam theories which are effective for different types of beam sections are known. Many refinements of original EBT kinematics have been proposed. Amongst these, the one attributed to Timoshenko in which transverse shear deformations are included should be mentioned. The other refined theories mentioned herein are those by Vlasov and by Wagner, both of which lead to improved strain/stress field descriptions.

Over the last few decades, computational methods, in particular the finite element method, have made the use of classical beam theories much more successful and attractive. The possibility of solving complex framed structures with very different boundary conditions (mechanical and geometrical) has made it possible to analyze many complex problems involving thousands of degrees of freedom (DOFs) with acceptable accuracy. However, the difficulty of obtaining a complete stress/strain field in those sections with complex geometries or thin walls still remains an open question which can be addressed by refined and advanced beam theories.

During the last decade, the first author of this book proposed the Carrera Unified Formulation (CUF), which was first applied to plates and shells and then recently extended to beams. The CUF permits one to develop a large number of beam theories with a variable number of displacement unknowns by means of a concise notation and by referring to a few fundamental nuclei. Higher-order beam theories can easily be implemented on the basis of the CUF, and the accuracy of a large variety of beam theories can be established in a hierarchical and/or axiomatic vs. asymptotic sense. A modern form of beam theories can therefore be constructed in a hierarchical manner. The number of unknown variables is a free parameter of the problem. A 3D stress/strain field can be obtained by an appropriate choice of these variables for any type of beam problem: compact sections, thin-walled sections, bending, torsion, shear, localized loadings, static and dynamic problems.

This book details classical and modern beam theories. Accuracy of the known theories is established by using the modern technique in the CUF. Various beam problems, in particular beam sections from civil to aerospace applications (wing airfoils), are considered in static and dynamic problems. Numerical results are obtained using the MUL2 software, which is available on the web site www.mul2.com.

www.wiley.com/go/carrera

Introduction

A brief introduction to the contents of the book is given here together with an overview of the milestone contributions to beam structure analysis.

Why another book on beams?

There is no need for another book on beam theories. Many books are, in fact, available, which have been written by some of the most eminent and talented scientists in the theory of elasticity and structures. It would be extremely difficult to write a better book. So, *why a new book on beam theories*? The reason is the following: this book presents a method to deal with beam theories that has never been considered before. As will be explained in the following chapters, the method introduced by the first author over the last decade for plates and shells is applied here to beams to build a large class of 1D (beam) hierarchical (variable kinematic) theories, which are based on automatic techniques to build governing equations and/or finite element matrices. The resulting theories permit one to deal with any section geometries subjected to any loading conditions and, at the same time, to reach quasi-3D solutions. Such results make the present book unique.

Review of historical contributions

Beam theories are extensively used to analyze the structural behavior of slender bodies, such as columns, arches, blades, aircraft wings, and bridges. The main advantage of beam models is that they reduce the 3D problem to a set of variables that only depends on the beam-axis coordinate. The 1D structural elements obtained are simpler and computationally more efficient than 2D (plate/shell) and 3D (solid) elements. This feature makes beam theories very attractive for the static and dynamic analysis of structures.

The classical, most frequently employed theories are those by Euler–Bernoulli (Bernoulli, 1751; Euler, 1744), de Saint-Venant (1856a,b), and Timoshenko (1921,

1922). The first two do not account for transverse shear deformations. The Timoshenko model considers a uniform shear distribution along the cross-section of the beam. A comprehensive comparison of Euler–Bernoulli and Timoshenko theories was made by Mucichescu (1984). However, none of these theories can detect non-classical effects such as warping, out- and in-plane deformations, torsion–bending coupling, or localized boundary conditions, whether geometrical or mechanical. These effects are usually due to small slenderness ratios, thin walls, and the anisotropy of the materials.

Many methods have been proposed to overcome the limitations of classical theories and to allow the application of 1D models to any geometry or boundary condition. Many examples of these models can be found in many well-known books on the theory of elasticity, for example, the book by Novozhilov (1961). Recent developments in beam models have been obtained by means of different approaches: the introduction of shear correction factors, the use of warping functions based on the de Saint-Venant's solution, the variational asymptotic solution (VABS), generalized beam theories (GBTs), and higher-order beam models. Some of the most relevant contributions are discussed below.

A considerable amount of work has been done to try to improve the global response of classical beam theories through the use of appropriate shear correction factors, as in the books by Timoshenko and Goodier (1970) and by Sokolnikoff (1956). Amongst the many available articles on this issue, the papers by Cowper (1966), Krishna Murty (1985), Pai and Schulz (1999), and Mechab *et al.* (2008) are of particular interest. An extensive effort was made by Gruttmann and his co-workers (Gruttmann *et al.*, 1999; Gruttmann and Wagner, 2001; Wagner and Gruttmann, 2002) to compute shear correction factors for several structural cases: torsional and flexural shearing stresses in prismatic beams; arbitrary shaped cross-sections; wide, thin-walled, and bridge-like structures.

El Fatmi (El Fatmi, 2002, 2007a,b,c; El Fatmi and Zenzri, 2004) introduced improvements to the displacement models over the beam section by introducing a warping function, ϕ, to enhance the description of the normal and shear stress of the beam. End-effects due to boundary conditions have been investigated by means of this model, as in the work by Krayterman and Krayterman (1987).

The de Saint-Venant solution has been the theoretical base of many advanced beam models. The 3D elasticity equations were reduced to beam-like structures by Ladevéze and his co-workers (Ladéveze and Simmonds, 1996, 1998; Ladéveze *et al.*, 2004). The resulting solution was modeled as the sum of a de Saint-Venant part and a residual part and applied to high-aspect-ratio beams with thin-walled sections. Other beam theories have been based on the displacement field proposed by Ieşan (1986) and solved by means of a semi-analytical finite element by Dong and his co-workers (Dong *et al.*, 2001; Kosmatka *et al.*, 2001; Lin *et al.*, 2001; Lin and Dong, 2006).

Asymptotic-type expansions have been proposed by Berdichevsky *et al.* (1992) on the basis of variational methods. This work represents the starting point of an alternative approach to constructing refined beam theories where a characteristic parameter (e.g., the cross-sectional thickness of a beam) is exploited to build an asymptotic series. Those terms that exhibit the same order of magnitude as the parameter when it vanishes are retained. Some valuable contributions on asymptotic methods are those related to VABS models built by Volovoi *et al.* (1999), Volovoi and Hodges (2000), Popescu and Hodges (2000), Yu *et al.* (2002a,b) and Yu and Hodges (2004, 2005).

GBTs have been derived from Schardt's work (Schardt, 1966, 1989, 1994). GBTs enhance classical theories by exploiting piecewise beam descriptions of thin-walled sections. GBT has been extensively employed and extended, in various forms, by Silvetre and Camotim and their co-workers (Dinis *et al.*, 2006; Silvestre, 2002, 2003, 2007; Silvestre and Camotim, 2002). Many other higher-order theories which are based on enhanced displacement fields over the beam cross-section have been introduced to include non-classical effects. Some considerations of higher-order beam theories were made by Washizu (1968). An advanced model was proposed by Kanok-Nukulchai and Shik Shin (1984); these authors improved classical finite beam elements by introducing new degrees of freedom to describe cross-section behavior. Other refined beam models can be found in the excellent review by Kapania and Raciti (1989a,b) which focused on bending, vibration, wave propagations, buckling, and post-buckling. Aeroelastic problems of thin-walled structures were examined by means of higher-order beams by Librescu and Song (1992) and Qin and Librescu (2002).

The aforementioned literature overview clearly shows the interest in further developments of refined theories for beams.

Classical and modern approaches: variational methods and CUF

This book focuses on refined theories, with only generalized displacement variables, for the static and dynamic analysis of 1D structures, "beams," with compact and thin-walled sections. Higher-order models are obtained in the framework of the CUF. This formulation was developed over the last decade for plate/shell models (Carrera, 1995, 2002, 2003; Carrera *et al.*, 2008) and it has recently been extended to beam modeling (Carrera and Giunta, 2010). The present formulation has been exploited for the static analysis of compact and thin-walled structures (Carrera *et al.*, 2010a). Free-vibration analyses have been carried out on hollow cylindrical and wing models (Carrera *et al.*, 2011, 2011). A beam model with only displacement degrees of freedom has been developed (Carrera and Petrolo, 2010) and asymptotic-like results were obtained in Carrera and Petrolo (2011).

CUF is a hierarchical formulation which considers the order of the model as a free parameter (i.e., as input) of the analysis; in other words, refined models are obtained with no need for ad hoc formulations. Beam theories are obtained on the basis of Taylor-type expansions. Euler–Bernoulli and Timoshenko beam theories are obtained as particular cases. The finite element method is used to handle arbitrary geometries as well as geometrical and loading conditions.

Outline of the contents

A brief description of the book's layout is given here to provide a brief overview of what will be discussed. Chapter 1 presents the basic equations that the structural analysis is based on: equilibrium equations, strain–displacement geometrical relations, and constitutive equations. The principle of virtual displacements is also introduced in strong and weak forms.

Chapter 2 focuses on the description of classical beam theories: namely, the Euler–Bernoulli and Timoshenko models. The kinematics model of these theories is introduced and then strains, stresses, stress resultants, and elastica equations are derived. Numerical examples are given in order to highlight the differences between these two models.

The first refined model of this book is given in Chapter 3, where the complete linear expansion case is presented. Particular attention is given to the importance of the in-plane stretching terms that characterize this model. Examples are provided in order to underline the importance of these terms and the ineffectiveness of classical models to deal with in-plane stretching.

A first attempt toward the construction of a unified theory is made in Chapter 4, where the aforementioned classical and linear models are represented in a unified manner. The presented procedure represents the first fundamental step in obtaining the Carrera Unified Formulation (CUF). The chapter closes with a discussion on the Poisson locking phenomenon and its correction.

The CUF is introduced in Chapter 5. A detailed description of all the keypoints related to the CUF is provided. Strong and weak forms are provided. All the fundamental matrices are derived by means of the fundamental nuclei assembly system, which represents the core of the formulation.

Chapter 6 presents a number of higher-order theories, in terms of displacement and strain fields. Formulas for obtaining any-order models are provided in order to stress the hierarchical capabilities of the CUF.

A wide and comprehensive implementation guideline report is provided in Chapter 7 where all the main issues related to the finite element implementation of the CUF are addressed, including the assembly procedure, numerical integrations, and the main pre- and post-processing steps. A large number of numerical benchmarks are given for comparison purposes. Several chapters that deal with applications of CUF beam models then follow.

Chapter 8 presents the so-called shell-like capabilities of the model. Several structural problems are considered with particular attention being paid to thin-walled structures and point loads. Comparisons to shell and solid models are provided in terms of both accuracy and computational costs.

A buckling analysis is carried out in Chapter 9, whereas Chapter 10 shows the extension of CUF models to FGM (Functionally Graded Materials) made structures.

Chapter 11 presents the Arlequin method and its application to CUF beam models. Structures are analyzed by means of multiple-order beam theories, that is, the order of the beam model is locally tuned in order to optimize the computational costs of the analysis.

An extensive analysis of the effectiveness of higher-order theories is carried out in Chapter 12, where the so-called axiomatic–asymptotic method is presented and exploited to build reduced refined models on the basis of a given accuracy which is given as an input of the analysis. The effect of different characteristic parameters, such as slenderness, boundary conditions, and output variables, is considered.

References

Berdichevsky VL, Armanios E, and Badir A 1992 Theory of anisotropic thin-walled closed-cross-section beams. *Composites Engineering*, **2**(5–7), 411–432.

Bernoulli D 1751 De vibrationibus et sono laminarum elasticarum. *In: Commentarii Academiae Scientiarum Imperialis Petropolitanae*, Petropoli.

Carrera E 1995 A class of two dimensional theories for multilayered plates analysis. *Atti Accademia delle Scienze di Torino, Memorie Scienze Fisiche*, **19-20** , 49–87.

Carrera E 2002 Theories and finite elements for multilayered, anisotropic, composite plates and shells. *Archives of Computational Methods in Engineering*, **9**(2), 87–140.

Carrera E 2003 Theories and finite elements for multilayered plates and shells: a unified compact formulation with numerical assessment and benchmarking. *Archives of Computational Methods in Engineering*, **10**(3), 216–296.

Carrera E, Brischetto S, and Robaldo A 2008 Variable kinematic model for the analysis of functionally graded material plates. *AIAA Journal*, **46**, 194–203.

Carrera E and Giunta G 2010 Refined beam theories based on a unified formulation. *International Journal of Applied Mechanics*, **2**(1), 117–143.

Carrera E, Giunta G, Nali P, and Petrolo M 2010a Refined beam elements with arbitrary cross-section geometries. *Computers & Structures*, **88**(5–6), 283–293. DOI: 10.1016/j.compstruc.2009.11.002.

Carrera E, Petrolo M, and Nali P 2011 Unified formulation applied to free vibrations finite element analysis of beams with arbitrary section. *Shock and Vibrations*, **18**(3), 485–502 DOI: 10.3233/SAV-2010-0528.

Carrera E, Petrolo M, and Varello A 2011 Advanced beam formulations for free vibration analysis of conventional and joined wings. *Journal of Aerospace Engineering*. In press. DOI: 10.1061/(ASCE)AS.1943-5525.0000130.

Carrera E and Petrolo M 2010 Refined beam elements with only displacement variables and plate/shell capabilities. Submitted.

Carrera E and Petrolo M 2011 On the effectiveness of higher-order terms in refined beam theories. *Journal of Applied Mechanics*, **78**. DOI: 10.1115/1.4002207.

Cowper GR 1966 The shear coefficient in Timoshenko's beam theory. *Journal of Applied Mechanics*, **33**(2), 335–340.

de Saint-Venant A 1856a Mémoire sur la flexion des prismes. *Journal de Mathématiques pures et appliqués*, **1**, 89–189.

de Saint-Venant A 1856b Mémoire sur la torsion des prismes. *Académie des Sciences de l'Institut Impérial de Frances*, **14**, 233–560.

Dinis P, Camotim D, and Silvestre N 2006 GBT formulation to analyse the buckling behaviour of thin-walled members with arbitrarily branched open cross-sections. *Thin-Walled Structures*, **44**(1), 20–38.

Dong SB, Kosmatka JB, and Lin HC 2001 On Saint-Venant's problem for an inhomogeneous, anisotropic cylinder-Part I: Methodology for Saint-Venant solutions. *Journal of Applied Mechanics*, **68**(3), 376–381.

El Fatmi R 2002 On the structural behavior and the Saint Venant solution in the exact beam theory: application to laminated composite beams. *Computers & Structures*, **80**(16–17), 1441–1456.

El Fatmi R 2007a Non-uniform warping including the effects of torsion and shear forces. Part I: A general beam theory. *International Journal of Solids and Structures*, **44**(18–19), 5912–5929.

El Fatmi R 2007b Non-uniform warping including the effects of torsion and shear forces. Part II: Analytical and numerical applications. *International Journal of Solids and Structures*, **44**(18–19), 5930–5952.

El Fatmi R 2007c A non-uniform warping theory for beams. *Comptes Rendus Mecanique*, **335**, 476–474.

El Fatmi R and Zenzri H 2004 A numerical method for the exact elastic beam theory: applications to homogeneous and composite beams. *International Journal of Solids and Structures*, **41**, 2521–2537.

Euler L 1744 De curvis elasticis. *In: Methodus Inveniendi Lineas Curvas Maximi Minimive Proprietate Gaudentes, Sive Solutio Problematis Isoperimetrici Lattissimo Sensu Accepti*. Bousquet.

Gruttmann F and Wagner W 2001 Shear correction factors in Timoshenko's beam theory for arbitrary shaped cross-sections. *Computational Mechanics*, **27**, 199–207.

Gruttmann F, Sauer R, and Wagner W 1999 Shear stresses in prismatic beams with arbitrary cross-sections. *International Journal for Numerical Methods in Engineering*, **45**, 865–889.

Ieşan 1986 On Saint-Venant's problem. *Archive for Rational Mechanics and Analysis*, **91**, 363–373.

Kanok-Nukulchai W and Shik Shin Y 1984 Versatile and improved higher-order beam elements. *Journal of Structural Engineering*, **110**, 2234–2249.

Kapania K and Raciti S 1989a Recent advances in analysis of laminated beams and plates, part I: Shear effects and buckling. *AIAA Journal*, **27**(7), 923–935.

Kapania K and Raciti S 1989b Recent advances in analysis of laminated beams and plates, part II: Vibrations and wave propagation. *AIAA Journal*, **27**(7), 935–946.

Kosmatka JB, Lin HC, and Dong SB 2001 On Saint-Venant's problem for an inhomogeneous, anisotropic cylinder-Part II: Cross-sectional properties. *Journal of Applied Mechanics*, **68**(3), 382–391.

Krayterman BL and Krayterman, AB 1987 Generalized nonuniform torsion of beams and frames. *Journal of Structural Engineering*, **113**, 1772–1787.

Krishna Murty AV 1985 On the shear deformation theory for dynamic analysis of beams. *Journal of Sound and Vibration*, **101**(1), 1–12.

Ladéveze P and Simmonds J 1996 De nouveaux concepts en théorie des poutres pour des charges et des géométries quelconques. *Comptes Rendus de l'Academie des Sciences. Paris*, **332**, 445–462.

Ladéveze P and Simmonds J 1998 New concepts for linear beam theory with arbitrary geometry and loading. *European Journal of Mechanics–A/Solids*, **17**(3), 377–402.

Ladéveze P, Sanchez P, and Simmonds J 2004 Beamlike (Saint-Venant) solutions for fully anisotropic elastic tubes of arbitrary closed cross section. *International Journal of Solids and Structures*, **41**(7), 1925–1944.

Librescu L and Song O 1992 On the static aeroelastic tailoring of composite aircraft swept wings modelled as thin-walled beam structures. *Composites Engineering*, **2**, 497–512.

Lin HC and Dong SB 2006 On the Almansi-Michell problems for an inhomogeneous, anisotropic cylinder. *Journal of Mechanics*, **22**(1), 51–57.

Lin HC, Dong SB, and Kosmatka JB 2001 On Saint-Venant's problem for an inhomogeneous, anisotropic cylinder-Part III: End effects. *Journal of Applied Mechanics*, **68**(3), 392–398.

Mechab I, Tounsi A, Benatta MA, and Bedia EA 2008 Deformation of short composite beam using refined theories. *Journal of Mathematical Analysis and Applications*, **346**, 468–479.

Mucichescu DT 1984 Bounds for stiffness of prismatic beams. *Journal of Structural Engineering*, **110**, 1410–1414.

VV Novozhilov 1961 Theory of elasticity Pergamon

Pai PF and Schulz MJ 1999 Shear correction factors and an energy consistent beam theory. *International Journal of Solids and Structures*, **36**, 1523–1540.

Popescu B and Hodges DH 2000 On asymptotically correct Timoshenko-like anisotropic beam theory. *International Journal of Solids and Structures*, **37**, 535–558.

Qin Z and Librescu L 2002 On a shear-deformable theory of anisotropic thin-walled beams: further contribution and validations. *Composite Structures*, **56**, 345–358.

Schardt R 1966 Eine Erweiterung der technischen Biegetheorie zur berechnung prismatischer Faltwerke. *Der Stahlbau*, **35**, 161–171.

Schardt R 1989 *Verallgemeinerte technische Biegetheorie*. Springer.

Schardt R 1994 Generalized beam theory–an adequate method for coupled stability problems. *Thin-Walled Structures*, **19**, 161–180.

Silvestre N 2002 Second-order generalised beam theory for arbitrary orthotropic materials. *Thin-Walled Structures*, **40**(9), 791–820.

Silvestre N 2003 GBT buckling analysis of pultruded FRP lipped channel members. *Computers & Structures*, **81**(18-19), 1889–1904.

Silvestre N 2007 Generalised beam theory to analyse the buckling behaviour of circular cylindrical shells and tubes. *Thin-Walled Structures*, **45**(2), 185–198.

Silvestre N and Camotim D 2002 First-order generalised beam theory for arbitrary orthotropic materials. *Thin-Walled Structures*, **40**(9), 791–820.

Sokolnikoff IS 1956 *Mathematical Theory of Elasticity*. McGraw-Hill.

Timoshenko SP 1921 On the corrections for shear of the differential equation for transverse vibrations of prismatic bars. *Philosophical Magazine*, **41**, 744–746.

Timoshenko SP 1922 On the transverse vibrations of bars of uniform cross section. *Philosophical Magazine*, **43**, 125–131.

Timoshenko SP and Goodier JN 1970 *Theory of elasticity*. McGraw-Hill.

Volovoi VV and Hodges DH 2000 Theory of anisotropic thin-walled beams. *Journal of Applied Mechanics*, **67**, 453–459.

Volovoi VV, Hodges DH, Berdichevsky VL, and Sutyrin VG 1999 Asymptotic theory for static behavior of elastic anisotropic I-beams. *International Journal of Solid Structures*, **36**, 1017–1043.

Wagner W and Gruttmann F 2002 A displacement method for the analysis of flexural shear stresses in thin-walled isotropic composite beams. *Computers & Structures*, **80**, 1843–1851.

Washizu, K 1968 *Variational methods in elasticity and plasticity*. Pergamon.

Yu W and Hodges DH 2004 Elasticity solutions versus asymptotic sectional analysis of homogeneous, isotropic, prismatic beams. *Journal of Applied Mechanics*, **71**, 15–23.

Yu W and Hodges DH 2005 Generalized Timoshenko theory of the variational asymptotic beam sectional analysis. *Journal of the American Helicopter Society*, **50**(1), 46–55.

Yu W, Hodges DH, Volovoi VV, and Cesnik CES 2002a On Timoshenko-like modeling of initially curved and twisted composite beams. *International Journal of Solids and Structures*, **39**, 5101–5121.

Yu W, Volovoi VV, Hodges DH, and Hong X 2002b Validation of the variational asymptotic beam sectional analysis (VABS). *AIAA Journal*, **40**, 2105–2113.

1

Fundamental equations of continuous deformable bodies

This chapter recalls the geometrical and constitutive equations of continuum structural mechanics in the linear case. Symbols and reference systems that will be used throughout the book are also introduced.

1.1 Displacement, strain, and stresses

Let us consider a continuous deformable body C with volume V in a three-orthogonal Cartesian system x, y, z as shown in Figure 1.1. The following relevant variables are introduced:

- Displacements, **u**, of a point P:

$$\mathbf{u} = \{\mathbf{u_x}, \quad \mathbf{u_y}, \quad \mathbf{u_z}\}^T$$

When the body C is subjected to mechanical and natural boundary conditions, the three continuous functions $u_x(x, y, z)$, $u_y(x, y, z)$, $u_z(x, y, z)$ give

Beam Structures: Classical and Advanced Theories, First Edition. Erasmo Carrera, Gaetano Giunta and Marco Petrolo.

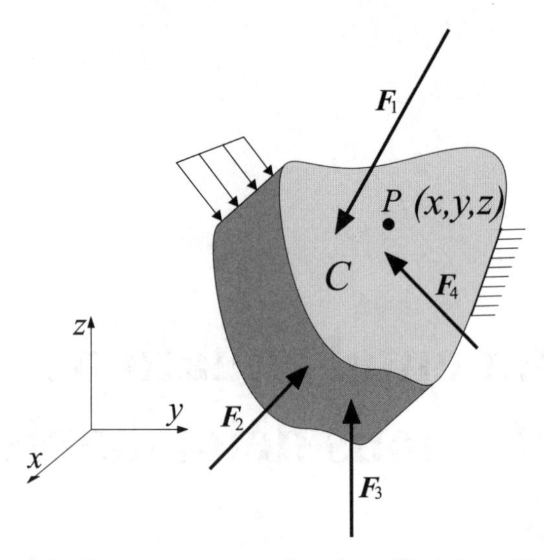

Figure 1.1 Geometry and notations for a 3D deformable body.

the deformed state of C. The calculation of such a deformed state remains the fundamental problem of 3D elasticity.

- Strains, $\boldsymbol{\epsilon}$:

$$\boldsymbol{\epsilon} = \{\epsilon_{xx}, \quad \epsilon_{yy}, \quad \epsilon_{zz}, \quad \epsilon_{xz}, \quad \epsilon_{yz}, \quad \epsilon_{xy}, \quad \epsilon_{zx}, \quad \epsilon_{yx}, \quad \epsilon_{yz}\}^T$$

These nine terms describe the deformed configuration in a three-orthogonal term of unit vectors ($\mathbf{i_x}$, $\mathbf{i_y}$, $\mathbf{i_z}$) in P. The axial deformations are

$$\epsilon_{xx}, \quad \epsilon_{yy}, \quad \epsilon_{zz}$$

Shear strains, which are symmetric, are

$$\epsilon_{xz} = \epsilon_{zx}, \quad \epsilon_{xy} = \epsilon_{yx}, \quad \epsilon_{yz} = \epsilon_{zy}$$

- Stresses, $\boldsymbol{\sigma}$, at a point P:

$$\boldsymbol{\sigma} = \{\sigma_{xx}, \quad \sigma_{yy}, \quad \sigma_{zz}, \quad \sigma_{xz}, \quad \sigma_{yz}, \quad \sigma_{xy}, \quad \sigma_{zx}, \quad \sigma_{yx}, \quad \sigma_{yz}\}^T$$

Axial stresses are

$$\sigma_{xx}, \quad \sigma_{yy}, \quad \sigma_{zz}$$

Shear stresses, which are symmetric, are

$$\sigma_{xz} = \sigma_{zx}, \quad \sigma_{xy} = \sigma_{yx}, \quad \sigma_{yz} = \sigma_{zy}$$

1.2 Equilibrium equations in terms of stress components and boundary conditions

At a generic point of the body C the following fundamental equilibrium equations hold:

$$\begin{cases} \dfrac{\partial \sigma_{xx}}{\partial x} + \dfrac{\partial \sigma_{yx}}{\partial y} + \dfrac{\partial \sigma_{zx}}{\partial z} = \mathbb{X} \\[2ex] \dfrac{\partial \sigma_{xy}}{\partial x} + \dfrac{\partial \sigma_{yy}}{\partial y} + \dfrac{\partial \sigma_{zy}}{\partial z} = \mathbb{Y} \\[2ex] \dfrac{\partial \sigma_{xz}}{\partial x} + \dfrac{\partial \sigma_{yz}}{\partial y} + \dfrac{\partial \sigma_{zz}}{\partial z} = \mathbb{Z} \end{cases} \tag{1.1}$$

where $\mathbb{X}$, $\mathbb{Y}$, and $\mathbb{Z}$ are the body forces per unit volume. On the boundary of the body C, the stress vector components at a generic point must fulfill the following conditions:

$$\begin{cases} \sigma_{xx}\, n_x + \sigma_{yx}\, n_y + \sigma_{zx}\, n_z = \mathbb{P}_x \\ \sigma_{xy}\, n_x + \sigma_{yy}\, n_y + \sigma_{zy}\, n_z = \mathbb{P}_y \\ \sigma_{xz}\, n_x + \sigma_{yz}\, n_y + \sigma_{zz}\, n_z = \mathbb{P}_z \end{cases} \tag{1.2}$$

where n_x, n_y, and n_z are the direction cosines, and $\mathbb{P}$ is the resultant of the external forces per unit area acting on the boundary of C. These are known as mechanical boundary conditions. Geometrical conditions can also be imposed stating that

$$\begin{cases} u_x = \mathbb{U}_x \\ u_y = \mathbb{U}_y \\ u_x = \mathbb{U}_z \end{cases} \tag{1.3}$$

where $\mathbb{U}_x$, $\mathbb{U}_y$, and $\mathbb{U}_z$ are imposed displacements on the outer surface of C.

1.3 Strain displacement relations

Strains are related to displacements by the following geometrical relations that are valid under the assumption of linearity:

$$\begin{cases} \epsilon_{xx} = \dfrac{\partial u_x}{\partial x} \\ \epsilon_{yy} = \dfrac{\partial u_y}{\partial y} \\ \epsilon_{zz} = \dfrac{\partial u_z}{\partial z} \\ \epsilon_{xy} = \dfrac{\partial u_x}{\partial y} + \dfrac{\partial u_y}{\partial x} \\ \epsilon_{zx} = \dfrac{\partial u_x}{\partial z} + \dfrac{\partial u_z}{\partial x} \\ \epsilon_{zy} = \dfrac{\partial u_y}{\partial z} + \dfrac{\partial u_z}{\partial y} \end{cases} \tag{1.4}$$

ε_{zy} and ε_{xy} are the engineering strain components which are usually denoted as γ_{zy} and γ_{xy}. In this book both symbols ε and γ are used to indicate the engineering strain components.

1.4 Constitutive relations: Hooke's law

Stress and strain components are not independent, a physical relation exists that depends on the used materials. In the linear elastic case, Hooke law states

$$\{\sigma\} = [\mathbf{C}]\ \{\epsilon\} \tag{1.5}$$

where the stiffness coefficients C_{ij} have been introduced. Compliances can be also used:

$$\{\epsilon\} = [\mathbf{S}]\ \{\sigma\} \tag{1.6}$$

For isotropic materials the explicit form of $\mathbf{C}$ is

$$\begin{bmatrix} C_{11} & C_{12} & C_{13} & 0 & 0 & 0 \\ C_{12} & C_{22} & C_{23} & 0 & 0 & 0 \\ C_{13} & C_{23} & C_{33} & 0 & 0 & 0 \\ 0 & 0 & 0 & C_{44} & 0 & 0 \\ 0 & 0 & 0 & 0 & C_{55} & 0 \\ 0 & 0 & 0 & 0 & 0 & C_{66} \end{bmatrix} \tag{1.7}$$

where

$$\begin{aligned} C_{11} = C_{22} = C_{33} &= \frac{E(1-\nu)}{(1+\nu)(1-2\nu)} \\ C_{12} = C_{13} = C_{23} &= \frac{\nu E}{(1+\nu)(1-2\nu)} \\ C_{44} = C_{55} = C_{66} &= G \end{aligned} \tag{1.8}$$

By referring to compliances one has

$$\mathbf{S} = \begin{bmatrix} S_{11} & S_{12} & S_{13} & 0 & 0 & 0 \\ S_{12} & S_{22} & S_{23} & 0 & 0 & 0 \\ S_{13} & S_{23} & S_{33} & 0 & 0 & 0 \\ 0 & 0 & 0 & S_{44} & 0 & 0 \\ 0 & 0 & 0 & 0 & S_{55} & 0 \\ 0 & 0 & 0 & 0 & 0 & S_{66} \end{bmatrix} \tag{1.9}$$

where

$$\begin{aligned} S_{11} = S_{22} = S_{33} &= 1/E \\ S_{12} = S_{23} = S_{13} &= -(\nu/E) \\ S_{44} = S_{55} = S_{66} &= 1/G \end{aligned} \tag{1.10}$$

E is Young's modulus, G is the shear modulus, and ν is the Poisson ratio, which are related by

$$G = \frac{E}{2\,(1+\nu)} \tag{1.11}$$

1.5 Displacement approach via principle of virtual displacements

The 3D equations written above are only part of the known equations of deformable bodies. Other variables could be introduced, such as stress functions as well as compatibility conditions on strains. Appropriate variational statements can be introduced to establish the variational form of the 3D equations, well-known examples being: the principle of virtual displacements, the principle of virtual forces, and mixed variational principles (such as the Hu–Washizu variational statement). These are not discussed in this book, details can be found in the book by Washizu (1968).

The main advantages of referring to variational statements are the following:

- Governing 3D equations can be written in both weak and strong form.
- Approximated theories can be built and related *variationally consistent* governing equations can be obtained.

The 3D equations can be formulated in different ways according to the choice made for the unknown variables, that is, displacement, stress, or mixed formulations can be used.

This book only refers to displacement unknowns. The related variational statement is the principle of virtual displacement (PVD), which states that the virtual variation of the internal work, δL_{int} has to be equal to the virtual variation of the external and inertial work, δL_{ext} and δL_{ine}:

$$\underbrace{\int_V \delta \epsilon^T \sigma dV}_{\delta L_{int}} = \delta L_{ext} + \delta L_{ine} \tag{1.12}$$

where V is the volume domain of the body.

In the displacement-formulated theory of structures, the displacement variables are expressed in terms of other displacement values or variables. If a 3D problem is considered, the generic displacement variable, s, will be written as

$$s(x, y, x) = F_\tau(x, y, z)S_\tau, \quad \tau = 1, M \tag{1.13}$$

where $F_\tau(x, y, z)$ are base functions exploited to approximate s above V; S_τ are the problem unknowns, which can be displacement values, their derivatives, or generic higher-order terms. The meaning of S_τ depends on the adopted expansion. PVD is used to obtain governing equations according to the S_τ variables.

The role of F_τ is related to the model to be approximated. In the case of plates, for instance, base functions are introduced along the thickness direction, see Figure 1.2:

$$F_{\tau_{plate}} = f(z) \tag{1.14}$$

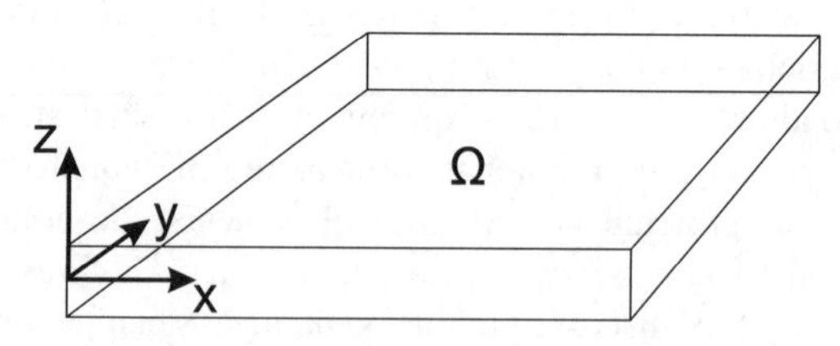

Figure 1.2 Plate domain and contour.

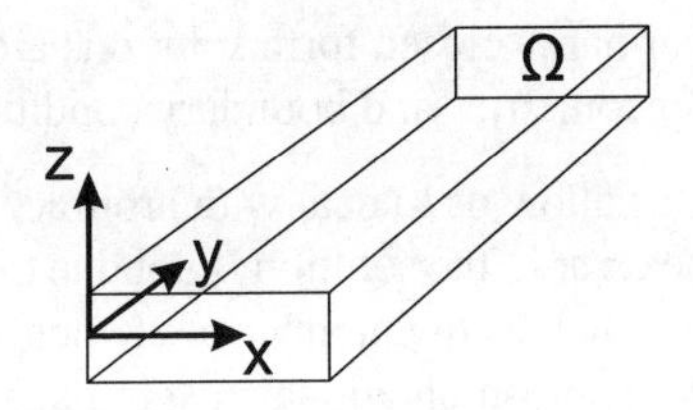

Figure 1.3 Plate domain and contour.

This means that approximations are introduced on the displacement field above the thickness of the plate. In the case of beams, base functions are introduced above the cross-section, Ω, see Figure 1.3:

$$F_{\tau_{beam}} = f(x, z) \tag{1.15}$$

The displacement field above the cross-section is then modeled by means of F_τ(x, z).

Governing equations can be formulated in strong and weak forms. Both formulations will be utilized in this book. Strong form solutions typically have two contributions:

- The contribution from the equilibrium at a point of the domain:

$$\delta S_s : K_{\tau,s} S_\tau = P_\tau \tag{1.16}$$

  where P_τ is the external load and $K_{\tau,s}$ the stiffness matrix.

- The contribution from boundary conditions:

$$\Pi_{\tau,s} S_\tau = \Pi_{\tau,s} \bar{S}_\tau \tag{1.17}$$

  where $\bar{S}_\tau$ is the imposed displacement at boundaries.

Weak form solutions in this book are obtained by means of finite elements (FEs). Only one governing equation is obtained in this case:

$$\delta \mathbf{q}_{\tau i}^T \mathbf{K}^{ij\tau s} \mathbf{q}_{sj} = P_{\tau i} \tag{1.18}$$

where i and j are indexes related to the interpolation polynomials named shape functions. Boundary conditions are imposed by acting on the stiffness matrix. The following considerations about strong and weak forms should be noted:

- Strong form solutions furnish exact solutions consistent with the approximations related to the structural model adopted. This means that the only error source is related to the model assumptions. The dimension of the matrices, that is, the computational cost, is given by the number of unknowns in the model. On the other hand, closed form solutions are usually available for a limited number of geometries and boundary conditions.

- Weak form solutions allow us to deal with arbitrary geometries and boundary conditions. The error in the solution is not due to the model assumptions alone, more factors, such as the number of elements used in the case of FE models, play a role. The computational cost is also related to the discretization refinement level: the finer the mesh, the larger the computational effort.

A detailed description of both formulations will be given in the following chapters of this book.

The keypoint in developing theories of structures consists of the appropriate choice of the unknown variables to trade off the complexity of the theory and its accuracy with respect to a known "exact solution," if any available. This book represents a contribution on the use of generalized expansions such as Equation 1.5 for beams to construct the governing equations by means of PVD. Hierarchical beam theories or, better, 1D models will be introduced for the analysis of any type of structure by referring to generalized displacement variables only.

Reference

Washizu, K 1968 Variational methods in elasticity and plasticity Pergamon.

2

The Euler–Bernoulli and Timoshenko theories

A beam is a structure whose axial extension, l, is predominant when compared to any other dimension orthogonal to it. The beam cross-section, Ω, is identified by intersecting the structure with planes that are orthogonal to its axis. Beam geometry, displacements, strains, and stresses are referred to an orthonormal Cartesian reference coordinate frame as shown in Figure 2.1; x and z are the in-plane coordinates lying on Ω and y is along the beam axis.

Unless stated differently, this reference system is adopted.

The mechanics of a beam under bending was first understood and described by Leonardo da Vinci as stated in Reti (1974) and Ballarini (2003). The Euler–Bernoulli (Euler, 1744) (EBBT) and the Timoshenko (1921, 1922) (TBT) models represent the classical beam theories. They are the reference models to analyze slender homogeneous structures under bending loads. This chapter is devoted to the description of both EBBT and TBT. Starting from the a priori hypotheses for the kinematics of a beam under bending, the displacement field, strains, stresses, and resulting forces will be derived. The limitations and the differences between the two models will be pointed out by means of some practical examples.

2.1 The Euler–Bernoulli model

The EBBT is derived from the following a priori assumptions, see Figure 2.2:

(I) the cross-section is rigid on its plane;

Beam Structures: Classical and Advanced Theories, First Edition. Erasmo Carrera, Gaetano Giunta and Marco Petrolo.

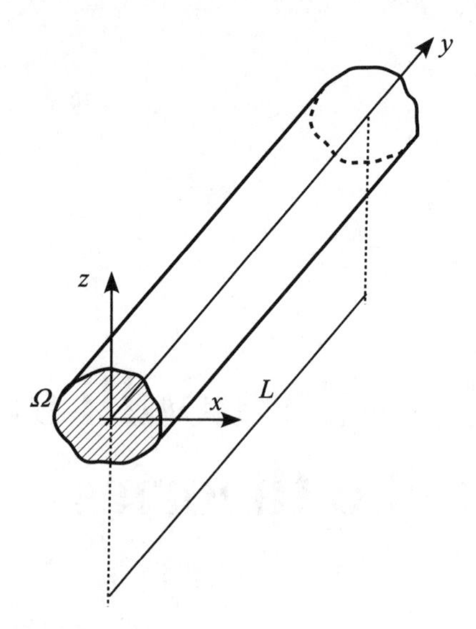

Figure 2.1 Beam reference coordinate frame.

(II) the cross-section rotates around a neutral surface remaining plane;

(III) the cross-section remains perpendicular to the neutral surface during deformation.

2.1.1 Displacement field

According to the first hypothesis, no in-plane deformations are accounted for and, therefore, the in-plane displacements u_x and u_z depend upon the axial

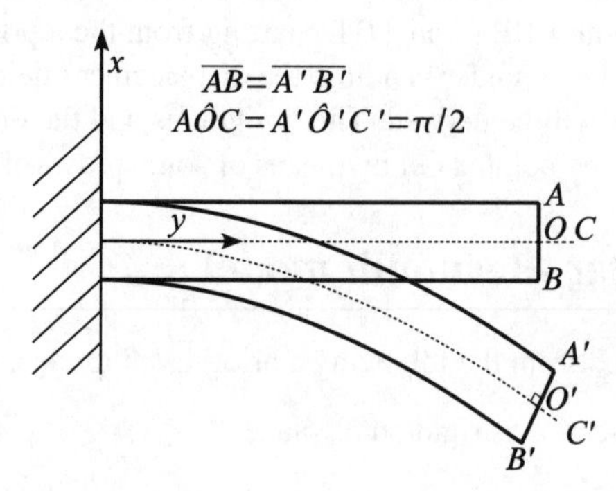

Figure 2.2 Bending of a beam according to the Euler–Bernoulli kinematic hypotheses.

coordinate y only:

$$\begin{cases} \epsilon_{xx} = \dfrac{\partial u_x}{\partial x} = 0 \\ \epsilon_{zz} = \dfrac{\partial u_z}{\partial z} = 0 \\ \gamma_{xz} = \dfrac{\partial u_x}{\partial z} + \dfrac{\partial u_z}{\partial x} = 0 \end{cases} \Rightarrow \begin{cases} u_x(x, y, z) = u_{x_1}(y) \\ u_z(x, y, z) = u_{z_1}(y) \end{cases} \tag{2.1}$$

On the basis of the second hypothesis, the out-of-plane or axial displacement u_y is linear versus the in-plane coordinates:

$$u_y(x, y, z) = u_{y_1}(y) + \phi_z(y)x + \phi_x(y)z \tag{2.2}$$

where ϕ_z and ϕ_x are the rotation angles along the z- and x-axis, respectively. ϕ_z is positive when, according to the "right-hand grip rule," the thumb points in the positive direction of the z-axis, whereas the thumb points in the negative direction of the x-axis for positive values of ϕ_x. On the basis of the third hypothesis and according to the definition of shear strains, shear deformations γ_{yz} and γ_{yx} are disregarded:

$$\gamma_{yz} = \gamma_{yx} = 0 \tag{2.3}$$

Equations 2.1, 2.2, and 2.3 allow the rotation angles to be obtained as functions of the derivatives of the in-plane displacements:

$$\begin{cases} \epsilon_{xy} = \dfrac{\partial u_y}{\partial x} + \dfrac{\partial u_x}{\partial y} = \phi_z + \dfrac{\partial u_{x_1}}{\partial y} = 0 \\ \epsilon_{yz} = \dfrac{\partial u_y}{\partial z} + \dfrac{\partial u_z}{\partial y} = \phi_x + \dfrac{\partial u_{z_1}}{\partial y} = 0 \end{cases} \Rightarrow \begin{cases} \phi_z = -\dfrac{\partial u_{x_1}}{\partial y} \\ \phi_x = -\dfrac{\partial u_{z_1}}{\partial y} \end{cases} \tag{2.4}$$

The displacement field of the EBBT is then

$$\begin{aligned} u_x &= u_{x_1} \\ u_y &= u_{y_1} - \frac{\partial u_{x_1}}{\partial y}x - \frac{\partial u_{z_1}}{\partial y}z \\ u_z &= u_{z_1} \end{aligned} \tag{2.5}$$

From a mathematical point of view, the EBBT displacement field can be seen as a Maclaurin-like series expansion in which a zero-order approximation is used for the in-plane components and an expansion order N equal to one is adopted for the axial displacement, and the relations amongst the unknowns have been derived from kinematic considerations. The EBBT presents three unknown variables.

2.1.2 Strains

According to the kinematic hypotheses, the EBBT accounts for the axial strain only. On the basis of its definition, and of the EBBT displacement field, ϵ_{yy} is

$$\epsilon_{yy} = \frac{\partial u_y}{\partial y} = \underbrace{\frac{\partial u_{y_1}}{\partial y}}_{k_y^y} \underbrace{- \frac{\partial^2 u_{x_1}}{\partial y^2}}_{k_{yy}^x} x \underbrace{- \frac{\partial^2 u_{z_1}}{\partial y^2}}_{k_{yy}^z} z = k_y^y + k_{yy}^x x + k_{yy}^z z \tag{2.6}$$

The term k_y^y has the physical meaning of membrane deformation, whereas k_{yy}^x and k_{yy}^x, being the second-order derivatives of the transverse displacements, represent the curvatures in the case of infinitesimal deformations and small rotations.

2.1.3 Stresses and stress resultants

The axial stress, σ_{yy}, is obtained from the axial strain by means of the reduced constitutive equations:

$$\sigma_{yy} = E\epsilon_{yy} = E\left(k_y^y + k_{yy}^x x + k_{yy}^z z\right) \tag{2.7}$$

The stress resultants are obtained by integrating the axial stress on the cross-section, see Figure 2.3:

(I) axial force $N(y)$:

$$\begin{aligned} N(y) &= \int_\Omega \sigma_{yy}\, d\Omega = \int_\Omega E\left(k_y^y + k_{yy}^x x + k_{yy}^z z\right) d\Omega \\ &= E\left(k_y^y \underbrace{\int_\Omega d\Omega}_{A} + k_{yy}^x \underbrace{\int_\Omega x\, d\Omega}_{S_x} + k_{yy}^z \underbrace{\int_\Omega z\, d\Omega}_{S_z}\right) \end{aligned} \tag{2.8}$$

(II) bending moment versus the z-axis $M_z(y)$:

$$\begin{aligned} M_z(y) &= \int_\Omega \sigma_{yy} z\, d\Omega = \int_\Omega E\left(k_y^y + k_{yy}^x x + k_{yy}^z z\right) x\, d\Omega \\ &= E\left(k_y^y \underbrace{\int_\Omega x\, d\Omega}_{S_x} + k_{yy}^x \underbrace{\int_\Omega x^2\, d\Omega}_{I_{zz}} + k_{yy}^z \underbrace{\int_\Omega xz\, d\Omega}_{I_{xz}}\right) \end{aligned} \tag{2.9}$$

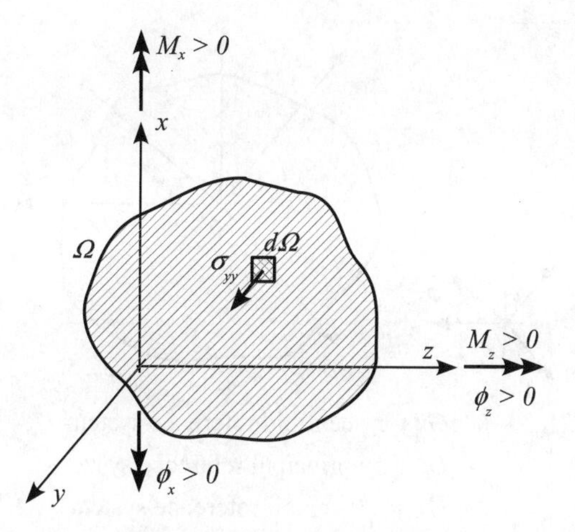

Figure 2.3 Stress resultants.

(III) bending moment versus the x-axis $M_x(y)$:

$$M_x(y) = -\int_{\Omega} \sigma_{yy} z \, d\Omega = -\int_{\Omega} E\left(k_y^y + k_{yy}^x x + k_{yy}^z z\right) z \, d\Omega$$

$$= -E\left(k_y^y \underbrace{\int_{\Omega} z \, d\Omega}_{S_z} + k_{yy}^x \underbrace{\int_{\Omega} xz \, d\Omega}_{I_{xz}} + k_{yy}^z \underbrace{\int_{\Omega} z^2 \, d\Omega}_{I_{xx}}\right) \tag{2.10}$$

These forces are statically equivalent to the stress distribution. The minus sign in Equation 2.10 is due to the fact that the bending moment M_x is positive when directed according to the positive direction of the x-axis and this yields a negative rotation angle ϕ_x as shown in Figure 2.3. A is the area of the cross-section (measure of the geometric entity Ω), S_x and S_z are its static momenta, and I_{xx}, I_{xz}, and I_{zz} are the cross-section momenta of inertia. Equations 2.8 to 2.9 can be written in matrix form:

$$\begin{Bmatrix} N \\ M_z \\ -M_x \end{Bmatrix} = E \begin{bmatrix} A & S_x & S_z \\ S_x & I_{zz} & I_{xz} \\ S_z & I_{xz} & I_{xx} \end{bmatrix} \begin{Bmatrix} k_1 \\ k_2 \\ k_3 \end{Bmatrix} \tag{2.11}$$

where

$$\mathbf{I}_{\Omega} = \begin{bmatrix} A & S_z & S_x \\ S_z & I_{zz} & I_{xz} \\ S_x & I_{xz} & I_{xx} \end{bmatrix} \tag{2.12}$$

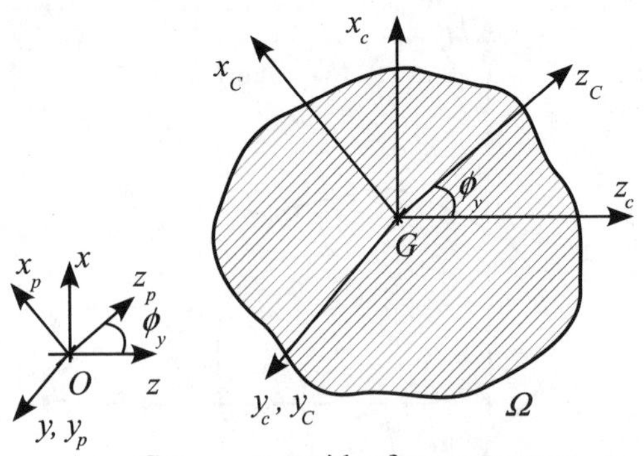

Figure 2.4 Cross-section reference systems.

is the fourth-order tensor of inertia. It should be noted that no assumptions for the Cartesian reference system have been introduced. The tensor of inertia and, as a consequence, Equations 2.11 can be simplified when a proper reference system is used, see Figure 2.4. The cross-section barycenter or centroid $G = (x_G, z_G)$ is the point at which, in an average sense, the area can be concentrated in

$$\begin{cases} \dfrac{1}{A}\displaystyle\int_\Omega (x - x_G)\, d\Omega = 0 \\ \dfrac{1}{A}\displaystyle\int_\Omega (z - z_G)\, d\Omega = 0 \end{cases} \Rightarrow \begin{cases} x_G = \dfrac{S_x}{A} \\ z_G = \dfrac{S_z}{A} \end{cases} \tag{2.13}$$

A centroidal reference system is centered on the cross-section barycenter. A reference system is called principal for a given cross-section when the product momentum computed according to that reference system is equal to zero. With reference to Figure 2.4, the angle ϕ_y between the x- and z-axes of a generic reference system and the corresponding ones in a principal reference system (x_C and z_C) can be obtained by putting $I_{x_Cz_C}$ equal to zero:

$$\begin{aligned} I_{x_Cz_C} &= \int_\Omega x_C z_C\, d\Omega \\ &= \int_\Omega (x\cos\phi_y - z\sin\phi_y)(x\sin\phi_y - z\cos\phi_y)\, d\Omega \\ &= \frac{1}{2}(I_{xx} + I_{zz})\sin 2\phi_y + I_{xz}\cos 2\phi_y = 0 \\ &\Rightarrow \phi_y = 2\tan^{-1}\left(-\frac{I_{xz}}{I_{xx} + I_{zz}}\right) \end{aligned} \tag{2.14}$$

A central reference system is a centroid principal reference system. In this reference system, the tensor of inertia is diagonal and Equations 2.11 are rewritten as follows:

$$\begin{Bmatrix} N_C \\ M_{z_C} \\ -M_{x_C} \end{Bmatrix} = E \begin{bmatrix} A & 0 & 0 \\ 0 & I_{z_C z_C} & 0 \\ 0 & 0 & I_{x_C x_C} \end{bmatrix} \begin{Bmatrix} k^{y_C}_{y_C} \\ k^{x_C}_{y_C y_C} \\ k^{z_C}_{y_C y_C} \end{Bmatrix} \tag{2.15}$$

Inversion of the previous equations yields the membrane deformation and curvatures, in the central reference system, in terms of the stress resultants:

$$\begin{aligned} k^{y_C}_{y_C} &= \frac{N_C}{EA} \\ k^{x_C}_{y_C y_C} &= \frac{M_{z_C}}{EI_{z_C z_C}} \\ k^{z_C}_{y_C y_C} &= -\frac{M_{x_C}}{EI_{x_C x_C}} \end{aligned} \tag{2.16}$$

The axial stress σ_{yy} can be also expressed in terms of the stress resultants (in the central reference system):

$$\sigma_{z_C z_C} = \frac{N_C}{A} + \frac{M_{z_C}}{I_{z_C z_C}} x - \frac{M_{x_C}}{I_{x_C x_C}} z \tag{2.17}$$

This last expression is also known as the Navier equation for the longitudinal stress.

2.1.4 Elastica

The vertical displacement and the rotation of the cross-section are given by a set of uncoupled differential equations named *elastica*:

$$\begin{cases} \dfrac{\partial^2 u_{x_{C1}}}{\partial y_C^2} = \dfrac{M_{z_C}}{EI_{z_C z_C}} \\[2ex] \dfrac{\partial^2 u_{z_{C1}}}{\partial y_C^2} = -\dfrac{M_{x_C}}{EI_{x_C x_C}} \\[2ex] \phi_{z_C} = -\dfrac{\partial u_{x_{C1}}}{\partial y_C} \\[2ex] \phi_{x_C} = -\dfrac{\partial u_{z_{C1}}}{\partial y_C} \end{cases} \tag{2.18}$$

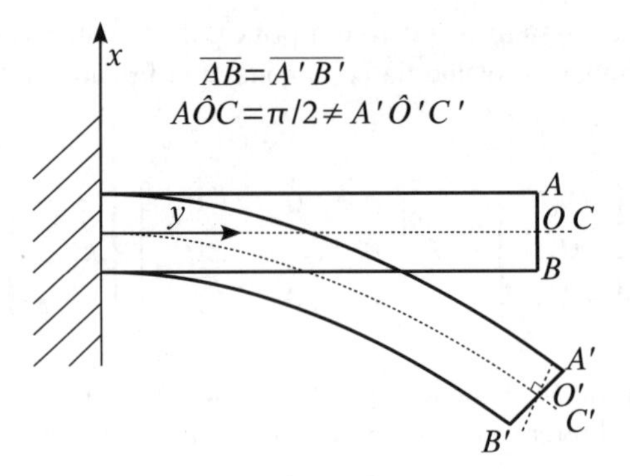

Figure 2.5 Bending of a beam according to Timoshenko's kinematic hypotheses.

2.2 The Timoshenko model

In the TBT the third kinematic a priori assumption of the EBBT is relaxed. The cross-section is still rigid on its plane, it rotates around a neutral surface remaining plane, but it is no longer constrained to remain perpendicular to it, see Figure 2.5. Shear deformations ϵ_{xy} and ϵ_{yz} are now accounted for.

2.2.1 Displacement field

According to the previous a priori kinematic assumptions, the displacement field of the TBT is

$$
\begin{aligned}
u_x(x, y, z) &= u_{x_1}(y) \\
u_y(x, y, z) &= u_{y_1}(y) + \phi_z(y)x + \phi_x(y)z \\
u_z(x, y, z) &= u_{z_1}(y)
\end{aligned}
\tag{2.19}
$$

2.2.2 Strains

The strain components are obtained by substituting the displacement field in Equations 2.19 into the geometrical relations in Equations 1.4. Only the non-null strain components are reported:

$$
\begin{aligned}
\epsilon_{yy} &= \frac{\partial u_y}{\partial y} = \frac{\partial u_{y_1}}{\partial y} + \frac{\partial \phi_z}{\partial y}x + \frac{\partial \phi_x}{\partial y}z \\
\gamma_{xy} &= \frac{\partial u_y}{\partial x} + \frac{\partial u_x}{\partial y} = \phi_z + \frac{\partial u_{x_1}}{\partial y} \\
\gamma_{yz} &= \frac{\partial u_y}{\partial z} + \frac{\partial u_z}{\partial y} = \phi_x + \frac{\partial u_{z_1}}{\partial y}
\end{aligned}
\tag{2.20}
$$

2.2.3 Stresses and stress resultants

The constitutive relations are used to obtain the axial stress and the shear stress components:

$$\begin{aligned}
\sigma_{yy} &= E\epsilon_{yy} = E\left(\frac{\partial u_{y_1}}{\partial y} + \frac{\partial \phi_z}{\partial y}x + \frac{\partial \phi_x}{\partial y}z\right)\\
\sigma_{xy} &= \kappa G\left(\phi_z + \frac{\partial u_{x_1}}{\partial y}\right)\\
\sigma_{yz} &= \kappa G\left(\phi_x + \frac{\partial u_{z_1}}{\partial y}\right)
\end{aligned} \tag{2.21}$$

where κ is the shear correction factor. The shear predicted by the TBT should be corrected since the model yields a constant value above the cross-section, whereas it is at least parabolic in order to satisfy the stress-free boundary conditions on the unloaded edges of the cross-section. The shear correction factor is mainly related to the cross-section geometry. In the literature there are many methods to compute κ, see, for instance, Timoshenko (1921), Cowper (1966), Pai and Schulz (1999), Gruttmann *et al.* (1999), and Gruttmann and Wagner (2001). A discussion on the shear correction factor is beyond the scope of this book. It will be shown that higher-order models yield shear stresses that are compliant with the mechanical boundary conditions.

The stress resultants are obtained by integrating the axial stress on the cross-section:

(I) axial force N:

$$N = \int_{\Omega} \sigma_{yy}\, d\Omega = E\int_{\Omega}\left(\frac{\partial u_{y_1}}{\partial y} + \frac{\partial \phi_z}{\partial y}x + \frac{\partial \phi_x}{\partial y}z\right) d\Omega \tag{2.22}$$

(II) bending moment versus the z-axis M_z:

$$M_z = \int_{\Omega} \sigma_{yy} z\, d\Omega = E\int_{\Omega}\left(\frac{\partial u_{y_1}}{\partial y}z + \frac{\partial \phi_z}{\partial y}xz + \frac{\partial \phi_x}{\partial y}z^2\right) d\Omega \tag{2.23}$$

(III) bending moment versus the x-axis M_x:

$$M_x = -\int_{\Omega} \sigma_{yy} z\, d\Omega = -E\int_{\Omega}\left(\frac{\partial u_{y_1}}{\partial y}x + \frac{\partial \phi_z}{\partial y}x^2 + \frac{\partial \phi_x}{\partial y}xz\right) d\Omega \tag{2.24}$$

(IV) shear force along the x-axis V_x:

$$V_x = \int_{\Omega} \sigma_{xy}\, d\Omega = \int_{\Omega} \kappa G\left(\phi_z + \frac{\partial u_{x_1}}{\partial y}\right) d\Omega = \kappa G\left(\phi_z + \frac{\partial u_{x_1}}{\partial y}\right) A \tag{2.25}$$

(V) shear force along the z-axis V_z:

$$V_z = \int_{\Omega} \sigma_{yz} \, d\Omega = \int_{\Omega} \kappa G \left(\phi_x + \frac{\partial u_{z_1}}{\partial y} \right) d\Omega = \kappa G \left(\phi_x + \frac{\partial u_{z_1}}{\partial y} \right) A \tag{2.26}$$

In a central reference system, the axial force and the bending moments become

$$\begin{aligned}
N_C &= \int_{\Omega} EA \frac{\partial u_{y_{C1}}}{\partial y_C} \, d\Omega = EAk_1 \\
M_{z_C} &= EI_{z_C z_C} \frac{\partial \phi_{z_C}}{\partial y_C} \\
M_{x_C} &= -EI_{x_C x_C} \frac{\partial \phi_{x_C}}{\partial y_C}
\end{aligned} \tag{2.27}$$

2.2.4 Elastica

The elastica equations are obtained by solving the following set of coupled differential equations:

$$\begin{cases}
\dfrac{\partial \phi_{z_C}}{\partial y_C} = \dfrac{M_{z_C}}{EI_{z_C z_C}} \\
\dfrac{\partial \phi_{x_c}}{\partial y_C} = -\dfrac{M_{x_C x_C}}{EI_{x_C x_C}} \\
\dfrac{\partial u_{x_{C1}}}{\partial y_C} = \dfrac{V_{x_C}}{\kappa G A} - \phi_{z_C} \\
\dfrac{\partial u_{z_{C1}}}{\partial y_C} = \dfrac{V_{z_C}}{\kappa G A} - \phi_{x_C}
\end{cases} \tag{2.28}$$

2.3 Bending of a cantilever beam: EBBT and TBT solutions

A cantilever beam is now considered to highlight the differences between EBBT and TBT solutions. Figure 2.6 shows a graphic description of the structure that is loaded by a vertical force applied at the free tip.

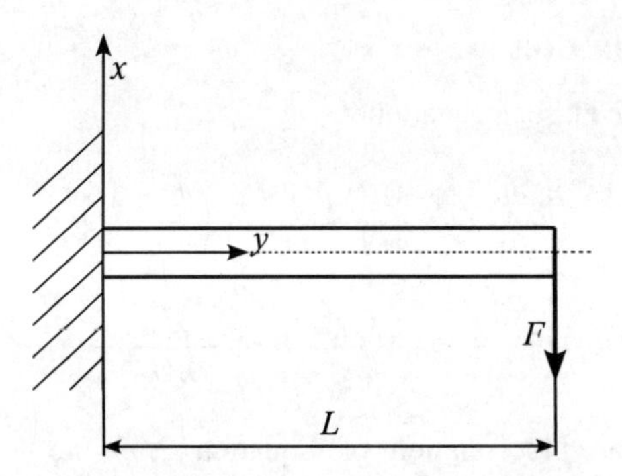

Figure 2.6 Cantilever beam bent by a vertical force.

2.3.1 EBBT solution

Let us consider the elastica equation

$$\frac{\partial^2 u_{x_1}}{\partial y^2} = \frac{M_z(y)}{EI_z} = -\frac{FL}{EI_z}\left(\frac{L-y}{L}\right) \tag{2.29}$$

By integrating Equation 2.29 the following relations are obtained:

$$\begin{aligned} \frac{\partial u_{x_1}}{\partial y} &= -\frac{FL}{EI_z}\left(y - \frac{y^2}{2L}\right) + C_1 \\ u_{x_1} &= -\frac{FL}{EI_z}\left(\frac{y^2}{L} - \frac{y^3}{6L}\right) + C_1 y + C_2 \end{aligned} \tag{2.30}$$

The following boundary conditions are applied in order to evaluate C_1 and C_2:

$$\begin{aligned} \left.\frac{\partial u_{x_1}}{\partial y}\right|_{y=0} &= 0 \Rightarrow C_1 = 0 \\ \left.u_{x_1}\right|_{y=1} &= 0 \Rightarrow C_2 = 0 \end{aligned} \tag{2.31}$$

The value of the maximum vertical displacement is then given by

$$\left.u_{x_1}\right|_{y=1} = -\frac{FL^3}{3EI_z} \tag{2.32}$$

2.3.2 TBT solution

Let us consider the elastica equation

$$\begin{cases} \dfrac{\partial \theta}{\partial y} = -\dfrac{M_z(y)}{EI_z} = \dfrac{FL}{EI_z}\left(\dfrac{L-y}{L}\right) \\ \dfrac{\partial u_{x_1}}{\partial y} = \dfrac{V_x(y)}{\kappa GA} - \theta = -\dfrac{F}{\kappa GA} - \theta \end{cases} \tag{2.33}$$

By integrating the first relation of Equation 2.33 the following relations are obtained:

$$\begin{cases} \theta = \dfrac{FL}{EI_z}\left(y - \dfrac{y^2}{2L}\right) + C_1 \\ \dfrac{\partial u_{x_1}}{\partial y} = -\dfrac{F}{\kappa GA} - \dfrac{FL}{EI_z}\left(y - \dfrac{y^2}{2L}\right) - C_1 \end{cases} \tag{2.34}$$

A further integration leads to

$$\begin{cases} \theta = \dfrac{FL}{EI_z}\left(y - \dfrac{y^2}{2L}\right) + C_1 \\ u_{x_1} = -\dfrac{Fy}{\kappa GA} - \dfrac{FL}{EI_z}\left(\dfrac{y^2}{2L} - \dfrac{y^3}{6L}\right) - C_1 y + C_2 \end{cases} \tag{2.35}$$

The following boundary conditions are applied in order to evaluate C_1 and C_2:

$$\begin{aligned} \theta|_{y=0} &= 0 \Rightarrow C_1 = 0 \\ u_{x_1}\big|_{y=0} &= 0 \Rightarrow C_2 = 0 \end{aligned} \tag{2.36}$$

The value of the maximum vertical displacement is then given by

$$u_{x_1}\big|_{y=0} = -\underbrace{\frac{FL}{\kappa GA}}_{\text{TBT, shear contribution}} - \underbrace{\frac{FL^3}{3EI_z}}_{\text{EBBT, bending contribution}} \tag{2.37}$$

By comparing this last equation to Equation 2.32, it is clear that the TBT solution provides the EBBT solution enhanced by the shear deformation contribution. Numerical evaluations of TBT and EBBT models are provided in Example 2.3.1 to highlight the importance of the shear deformation and TBT in the case of short beams. It has also to be emphasized that the TBT can assume a more significant role in the study of orthotropic and composite materials which generally have E/G ratios much larger than those of isotropic materials, that is, larger shear deformability (Kapania and Raciti, 1989). Similar considerations are still valid when the classical laminated theory (CLT) and the first order shear deformability theory (FSDT) for plates are considered (Carrera and Petrolo, 2010).

Example 2.3.1 *Let us consider the beam in Figure 2.6 to evaluate the maximum displacement given by the EBBT and TBT for different span length values. Let us assume* $E = 75$ GPa, $\nu = 0.33$, $F = 1$ N. *A square cross-section is considered with* $b = 0.01$ m, *the shear correction factor,* κ, *is assumed equal to* 5/6, *and the two span lengths to be considered,* L_1 *and* L_2, *are equal to* 1 *and* 0.1 m. *A slender and a moderately thick beam are considered. The following parameters are computed first:*

$$G = \frac{E}{2(1+\nu)} = 28.195 \text{ GPa}$$

$$A = b^2 = 0.0001 \text{ m}^2$$

$$I_z = \frac{b^4}{12} = 8.333 \times 10^{-10} \text{ m}^4$$

The slender beam case is considered first:

$$u_{x_{EBBT}} = -\frac{FL^3}{3EI_z} = 5.333 \times 10^{-3} \text{ m}$$

$$u_{x_{TBT}} = -\frac{FL}{\kappa GA} - \frac{FL^3}{3EI_z} = 5.334 \times 10^{-3} \text{ m}$$

The percentage variation is given by

$$\frac{u_{x_{TBT}} - u_{x_{EBBT}}}{u_{x_{TBT}}} \times 100 = 0.008\%$$

The moderately thick beam is then considered:

$$u_{x_{EBBT}} = -\frac{FL^3}{3EI_z} = 5.333 \times 10^{-6} \text{ m}$$

$$u_{x_{TBT}} = -\frac{FL}{\kappa GA} - \frac{FL^3}{3EI_z} = 5.376 \times 10^{-3} \text{ m}$$

The percentage variation is given by

$$\frac{u_{x_{TBT}} - u_{x_{EBBT}}}{u_{x_{TBT}}} \times 100 = 0.792\%$$

This simple example shows the importance of the shear deformation, of the TBT, in the case of short beams.

References

Ballarini R 2003 The Da Vinci-Euler-Bernoulli beam theory? *Mechanical Engineering Magazine*. Available Online, http://www.memagazine.org/contents/current/webonly/webex418.html.

Carrera E and Petrolo M 2010 Guidelines and recommendations to construct theories for metallic and composite plates. *AIAA Journal*, **48**(12), 2852–2866.

Cowper GR 1966 The shear coefficient in Timoshenko's beam theory. *Journal of Applied Mechanics*, **33**(2), 335–340.

Euler L 1744 De curvis elasticis. *In: Methodus Inveniendi Lineas Curvas Maximi Minimive Proprietate Gaudentes, Sive Solutio Problematis Isoperimetrici Lattissimo Sensu Accepti*. Bousquet.

Gruttmann F, Sauer R, and Wagner W 1999 Shear stresses in prismatic beams with arbitrary cross-sections. *International Journal for Numerical Methods in Engineering*, **45**, 865–889.

Gruttmann F and Wagner W 2001 Shear correction factors in Timoshenko's beam theory for arbitrary shaped cross-sections. *Computational Mechanics*, **27**, 199–207.

Kapania K and Raciti S 1989 Recent advances in analysis of laminated beams and plates, part I: Shear effects and buckling. *AIAA Journal*, **27**(7), 923–935.

Pai PF and Schulz MJ 1999 Shear correction factors and an energy consistent beam theory. *International Journal of Solids and Structures*, **36**, 1523–1540.

Reti, L 1974 *The unknown Leonardo*. McGraw-Hill.

Timoshenko SP 1921 On the corrections for shear of the differential equation for transverse vibrations of prismatic bars. *Philosophical Magazine*, **41**, 744–746.

Timoshenko SP 1922 On the transverse vibrations of bars of uniform cross section. *Philosophical Magazine*, **43**, 125–131.

3

A refined beam theory with in-plane stretching: the complete linear expansion case

This chapter presents a beam model that permits a linear variation of the three displacement components over the cross-section. The adoption of classical beam theories only allows one to deal with a linear distribution of the out-of-plane displacement component, while a constant in-plane stretching is provided. The complete linear expansion case (CLEC), overcomes this limitation by assuming a full linear expansion of the variables above the cross-section. After a theoretical description of CLEC, practical examples will be given to underline the importance of the linear stretching terms.

3.1 The CLEC displacement field

A linear expansion of Taylor-like basis functions is involved in the CLEC model

$$\begin{array}{rcl|l} u_x &=& u_{x_1} & +x\, u_{x_2} + z\, u_{x_3} \\ u_y &=& u_{y_1} & +x\, u_{y_2} + z\, u_{y_3} \\ u_z &=& \underbrace{u_{z_1}}_{N=0} & \underbrace{+x\, u_{z_2} + z\, u_{z_3}}_{N=1} \end{array} \tag{3.1}$$

Beam Structures: Classical and Advanced Theories, First Edition. Erasmo Carrera, Gaetano Giunta and Marco Petrolo.

The beam model given in Equation 3.1 has nine displacement variables: three constant ($N = 0$) and six linear ($N = 1$). Once the displacement variables are available, the strain components can be obtained straightforwardly:

$$
\begin{aligned}
\epsilon_{xx} &= \frac{\partial u_x}{\partial x} = u_{x_2} \\
\epsilon_{yy} &= \frac{\partial u_y}{\partial y} = u_{y_1,y} + x\, u_{y_2,y} + z\, u_{y_3,y} \\
\epsilon_{zz} &= \frac{\partial u_z}{\partial z} = u_{z_3} \\
\epsilon_{xy} &= \frac{\partial u_x}{\partial y} + \frac{\partial u_y}{\partial x} = u_{x_1,y} + x\, u_{x_2,y} + z\, u_{x_3,y} + u_{y_2} \\
\epsilon_{xz} &= \frac{\partial u_x}{\partial z} + \frac{\partial u_z}{\partial x} = u_{x_3} + u_{z_2} \\
\epsilon_{yz} &= \frac{\partial u_y}{\partial z} + \frac{\partial u_z}{\partial y} = u_{y_3} + u_{z_1,y} + x\, u_{z_2,y} + z\, u_{z_3,y}
\end{aligned}
\tag{3.2}
$$

The linear model leads to a constant distribution of ϵ_{xx}, ϵ_{zz}, and ϵ_{xz} above the cross-section, and a linear distribution of ϵ_{yy}, ϵ_{xy}, and ϵ_{yz}. Hooke's law can be used to compute the stresses.

3.2 The importance of linear stretching terms

The introduction of CLEC is motivated by the inadequacy of classical theories to model the in-plane stretching of the cross-section. This limitation cannot be neglected in many cases, especially when thin-walled structures are considered. However, the analysis of compact beams can also be inadequate if conducted with classical models. A simple example is given in this section to underline the importance of u_{x_2}, u_{x_3}, u_{z_2}, and u_{z_3}, that is, to motivate the need for CLEC.

Figure 3.1 shows a rectangular cross-section loaded by two point forces which provide a pure torsion state. The adoption of EBBT or TBT to analyze this problem

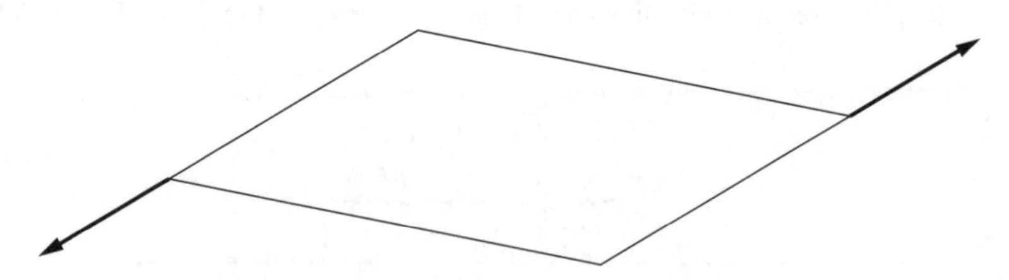

Figure 3.1 A square beam under torsion.

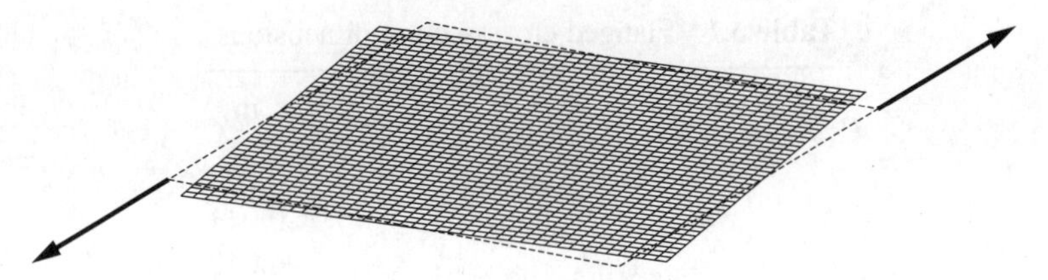

Figure 3.2 Deformed cross-section due to torsion via a linear model.

would provide null displacements since only constant in-plane translations can be considered. CLEC can provide a linear distribution of in-plane stretching that implies the deformed configuration shown in Figure 3.2.

In the case of a free-vibration analysis, similar considerations are still valid. CLEC can provide torsional frequencies, whereas EBBT and TBT cannot.

Example 3.2.1 *Let us consider a cantilevered beam of length* $L = 1$ m *having the cross-section shown in Figure 3.3; the dimensions of the cross-section are given in Table 3.1. An isotropic material is used with* $E = 70$ GPa *and* $\nu = 0.33$. *Two load cases are considered: in the first a force,* F_1, *is applied at* $[0, L, -h/2]$*; in the second one a force,* F_2, *is applied at* $[0.043, L, -h/2]$. *In both cases a*

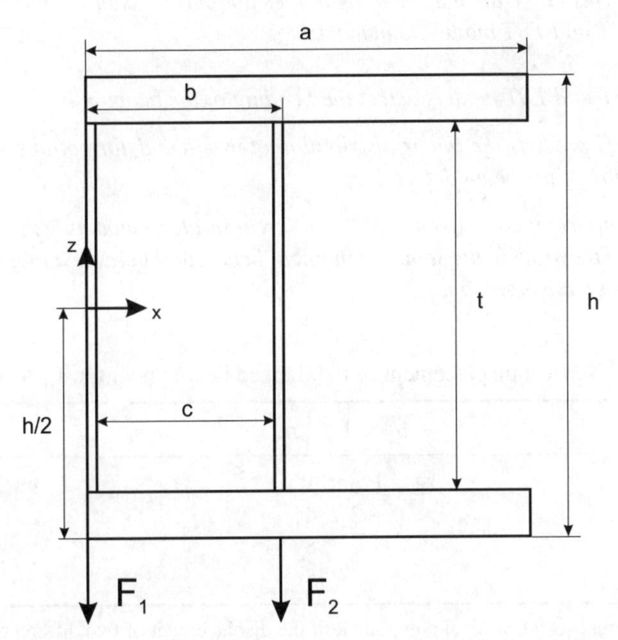

Figure 3.3 A flanged beam under a bending-torsion load.

Table 3.1 Flanged cross-section dimensions.

	m
a	0.100
b	0.044
c	0.040
h	0.100
t	0.080

This table presents the numerical values of the cross-section dimensions in Figure 3.3.

unitary load magnitude is considered. Analyses are conducted by means of EBBT, TBT, and CLEC based finite elements. Results are evaluated in terms of vertical displacements of two points: A, $[0, L, -h/2]$, and B, $[a, L, -h/2]$. Table 3.2 presents the results related to the first load case. The deformed free-tip cross-sections are shown in Figures 3.4 and 3.5 where EBBT, TBT, and CLEC solutions are compared. The role of CLEC is clear in detecting the torsion of the beam whereas classical models consider the bending behavior only. Table 3.3 presents the results related to the second load case when F_2 is applied. Classical models detect exactly the same solution of the first load case since the absence of linear in-plane terms in the displacement field does not permit us to include the torsion in the EBBT and TBT models. Summarizing:

- *EBBT and TBT models detect the bending of the beam only;*
- *CLEC predicts the linear distribution of in-plane deformations making the torsion of the beam detectable;*
- *an important consequence of the different in-plane modelings is that EBBT and TBT provide the same mechanical behavior wherever a load is applied above a cross-section.*

Table 3.2 Vertical displacement of the flanged beam under the F_1 load.

	$u_z \times 10^{-5}$ m	$u_z \times 10^{-5}$ m
	Point A	Point B
EBBT	−2.734	−2.734
TBT	−2.738	−2.738
CLEC	−2.745	−2.730

This table presents the vertical component of the displacement of two different points on the free-tip cross-section of the flanged cantilevered beam when F_1 is applied.

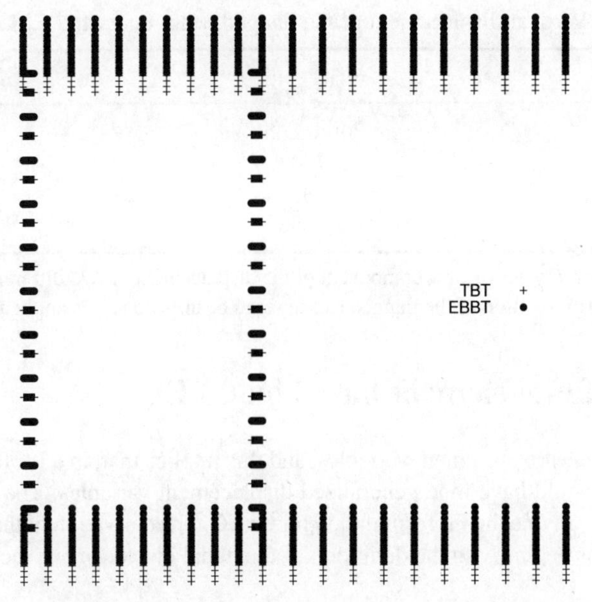

Figure 3.4 Deformed flanged cross-section via EBBT and EBT when F_1 is applied.

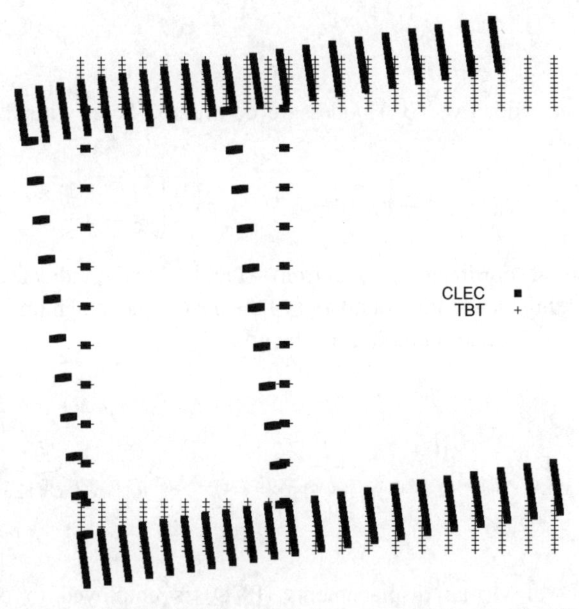

Figure 3.5 Deformed flanged cross-section via TBT and CLEC when F_1 is applied.

Table 3.3 Vertical displacement of the flanged beam under the F_1 load.

	$u_z \times 10^{-5}$ m	$u_z \times 10^{-5}$ m
	Point A	Point B
EBBT	−2.734	−2.734
TBT	−2.738	−2.738
CLEC	−2.738	−2.737

This table presents the vertical component of the displacement of two different points on the free-tip cross-section of the flanged cantilevered beam when F_2 is applied.

3.3 A finite element based on CLEC

CLEC has nine displacement variables, and this implies that, in a FE formulation, each node would have nine generalized displacement variables. The aim of this section is to provide the FE formulation for CLEC. The derivation of the governing FE equations begins with the definition of the nodal displacement vector

$$\mathbf{q}_i = \left\{ q_{u_{x_1}} \; q_{u_{y_1}} \; q_{u_{z_1}} \; q_{u_{x_2}} \; q_{u_{y_2}} \; q_{u_{z_2}} \; q_{u_{x_3}} \; q_{u_{y_3}} \; q_{u_{z_3}} \right\}^T \tag{3.3}$$

The displacement variables are interpolated along the beam axis by means of the shape functions, N_i:

$$\mathbf{u} = N_i \mathbf{q}_i \tag{3.4}$$

Beam elements with two (B2) nodes are considered here, with the following shape functions:

$$N_1 = \frac{1}{2}(1-r), \quad N_2 = \frac{1}{2}(1+r), \quad \begin{cases} r_1 = -1 \\ r_2 = +1 \end{cases} \tag{3.5}$$

where the natural coordinate, r, varies from -1 and $+1$ and r_i indicates the position of the node within the natural boundaries of the beam. The total number of degrees of freedom of the structural model will be given by

$$DOF = \underbrace{3 \times 3}_{\text{number of DOFs per node}} \times [(\underbrace{2}_{\text{number of nodes per element}} - 1) \times \underbrace{N_{BE}}_{\text{total number of beam elements}} + 1] \tag{3.6}$$

The principle of virtual displacements (PVD) is employed to compute the FE matrices

$$\delta L_{int} = \delta L_{ext} \tag{3.7}$$

where

$$\delta L_{int} = \int_V (\delta \epsilon_p^T \sigma_p + \delta \epsilon_n^T \sigma_n) dV \tag{3.8}$$

L_{int} stands for the strain energy, L_{ext} is the work of the external loadings, and δ stands for the virtual variation. Using Equation 3.5, a compact form of the virtual variation of the internal work can be obtained, as known from the FE procedure

$$\delta L_{int} = \delta \mathbf{q}_i^T \mathbf{K}^{ij} \mathbf{q}_j \tag{3.9}$$

where $\mathbf{K}^{ij}$ is the stiffness matrix. For a given i, j pair, the stiffness matrix has the form

$$\left[\begin{array}{ccc|ccc|ccc}
K_{xx}^{1,1} & K_{xy}^{1,1} & K_{xz}^{1,1} & K_{xx}^{1,x} & K_{xy}^{1,x} & K_{xz}^{1,x} & K_{xx}^{1,z} & K_{xy}^{1,z} & K_{xz}^{1,z} \\
K_{yx}^{1,1} & K_{yy}^{1,1} & K_{yz}^{1,1} & K_{yx}^{1,x} & K_{yy}^{1,x} & K_{yz}^{1,x} & K_{yx}^{1,z} & K_{yy}^{1,z} & K_{yz}^{1,z} \\
K_{zx}^{1,1} & K_{zy}^{1,1} & K_{zz}^{1,1} & K_{zx}^{1,x} & K_{zy}^{1,x} & K_{zz}^{1,x} & K_{zx}^{1,z} & K_{zy}^{1,z} & K_{zz}^{1,z} \\
\hline
K_{xx}^{x,1} & K_{xy}^{x,1} & K_{xz}^{x,1} & K_{xx}^{x,x} & K_{xy}^{x,x} & K_{xz}^{x,x} & K_{xx}^{x,z} & K_{xy}^{x,z} & K_{xz}^{x,z} \\
K_{yx}^{x,1} & K_{yy}^{x,1} & K_{yz}^{x,1} & K_{yx}^{x,x} & K_{yy}^{x,x} & K_{yz}^{x,x} & K_{yx}^{x,z} & K_{yy}^{x,z} & K_{yz}^{x,z} \\
K_{zx}^{x,1} & K_{zy}^{x,1} & K_{zz}^{x,1} & K_{zx}^{x,x} & K_{zy}^{x,x} & K_{zz}^{x,x} & K_{zx}^{x,z} & K_{zy}^{x,z} & K_{zz}^{x,z} \\
\hline
K_{xx}^{z,1} & K_{xy}^{z,1} & K_{xz}^{z,1} & K_{xx}^{z,x} & K_{xy}^{z,x} & K_{xz}^{z,x} & K_{xx}^{z,z} & K_{xy}^{z,z} & K_{xz}^{z,z} \\
K_{yx}^{z,1} & K_{yy}^{z,1} & K_{yz}^{z,1} & K_{yx}^{z,x} & K_{yy}^{z,x} & K_{yz}^{z,x} & K_{yx}^{z,z} & K_{yy}^{z,z} & K_{yz}^{z,z} \\
K_{zx}^{z,1} & K_{zy}^{z,1} & K_{zz}^{z,1} & K_{zx}^{z,x} & K_{zy}^{z,x} & K_{zz}^{z,x} & K_{zx}^{z,z} & K_{zy}^{z,z} & K_{zz}^{z,z}
\end{array}\right]_{ij} \tag{3.10}$$

where the superscripts indicate the expansion functions that are involved in each component of the stiffness matrix, that is, 1, x, or z. For the sake of clarity, the explicit expression for two components is reported hereafter:

$$\begin{aligned}
K_{xx}^{1,1} &= \tilde{C}_{44} \int_\Omega 1 \cdot 1 \, d\Omega \int_l N_{i,y} N_{j,y} dy \\
K_{yx}^{x,z} &= \tilde{C}_{23} \int_\Omega x \cdot \frac{\partial z}{\partial z} \, d\Omega \int_l N_{i,y} N_j dy \\
&= \tilde{C}_{23} \int_\Omega x \cdot 1 \, d\Omega \int_l N_{i,y} N_j dy
\end{aligned} \tag{3.11}$$

where Ω indicates the cross-sectional area and l the element length.

Example 3.3.1 *Let us consider a square beam element, its length being equal to L. A B2 element is used*

$$i, j = 1, 2 \Rightarrow N_1 = 1 - \frac{y}{L}, N_2 = \frac{y}{L}$$

and a linear theory, $N = 1$, having the following expansion functions:

$$1, x, \text{ and } z$$

The cross-section coordinates vary from $-a$ to $+a$ along both the x-direction and z-direction. The $K_{xx}^{x,x}$ component has to be computed at $i = 2$ and $j = 2$:

$$K_{xx}^{x,x} = \tilde{C}_{22} \int_{-a}^{+a} \int_{-a}^{+a} \frac{\partial x}{\partial x} \cdot \frac{\partial x}{\partial x} \, dxdz \int_0^L N_2 N_1 dy + \tilde{C}_{44} \int_{-a}^{+a} \int_{-a}^{+a} x \cdot x \, dxdz \int_0^L N_{2,_y} N_{1,_y} dy$$

by substituting the explicit expression of the functions the integrals become

$$K_{xx}^{2122} = \tilde{C}_{22} \int_{-a}^{+a} \int_{-a}^{+a} 1 \cdot 1 \, dxdz \int_0^L \frac{y}{L} \left(1 - \frac{y}{L}\right) dy + \tilde{C}_{44} \int_{-a}^{+a} \int_{-a}^{+a} x \cdot x \, dxdz \int_0^L \frac{1}{L} \left(-\frac{1}{L}\right) dy$$

The final result is

$$K_{xx}^{2122} = \frac{2}{3} \tilde{C}_{22} \, a^2 \, L - \frac{4}{3} \tilde{C}_{44} \frac{a^4}{L}$$

Now let us consider the normal strain components that have to be computed at a generic point with coordinates $[x_p, y_p, z_p]$. The strain components are given by

$$\epsilon_{xx} = \frac{\partial u_x}{\partial x} = \frac{\partial (u_{x_1})}{\partial x} N_i(y_p) \, q_{x_{1i}} + \frac{\partial (x \, u_{x_2})}{\partial x}\bigg|_{x=x_p} N_i(y_p) \, q_{x_{2i}} + \frac{\partial (z \, u_{x_3})}{\partial x}\bigg|_{z=z_p} N_i(y_p) \, q_{x_{3i}}$$

$$\epsilon_{yy} = \frac{\partial u_y}{\partial y} = u_{y_1} \frac{\partial N_i(y_p)}{\partial y}\bigg|_{y=y_p} q_{y_{1i}} + x_p \, u_{y_2} \frac{\partial N_i(y_p)}{\partial y}\bigg|_{y=y_p} q_{y_{2i}} + z_p \, u_{y_3} \frac{\partial N_i(y_p)}{\partial y}\bigg|_{y=y_p} q_{y_{3i}}$$

$$\epsilon_{zz} = \frac{\partial u_z}{\partial z} = \frac{\partial (u_{z_1})}{\partial x} N_i(y_p) \, q_{z_{1i}} + \frac{\partial (x \, z_{x_2})}{\partial x}\bigg|_{x=x_p} N_i(y_p) \, q_{z_{2i}} + \frac{\partial (z \, u_{z_3})}{\partial x}\bigg|_{z=z_p} N_i(y_p) \, q_{z_{3i}}$$

Thus

$$\epsilon_{xx}(x_p, y_p, z_p) = 1 \; N_i(y_p) \, q_{x_{2i}}$$

$$\epsilon_{yy}(x_p, y_p, z_p) = (1 \; q_{y_{1i}} + x_p \, q_{y_{2i}} + z_p \, q_{y_{3i}}) \frac{\partial N_i(y_p)}{\partial y}\bigg|_{y=y_p}$$

$$\epsilon_{zz}(x_p, y_p, z_p) = 1 \; N_i(y_p) \, q_{z_{3i}}$$

where $i = 1, 2$.

Further reading

Carrera E and Giunta G 2010 Refined beam theories based on a unified formulation. *International Journal of Applied Mechanics*, **2**(1), 117–143.

Carrera E and Petrolo M 2011 On the effectiveness of higher-order terms in refined beam theories. *Journal of Applied Mechanics*, **78**(2), DOI: 10.1115/1.4002207.

Carrera E, Giunta G, Nali P, and Petrolo M 2010 Refined beam elements with arbitrary cross-section geometries. *Computers & Structures*, **88**(5–6), 283–293. DOI: 10.1016/j.compstruc.2009.11.002.

Carrera E, Petrolo M, and Nali P 2011 Unified formulation applied to free vibrations finite element analysis of beams with arbitrary section. *Shock and Vibrations*, **18**(3), 485–502. DOI: 10.3233/SAV-2010-0528.

Kapania K and Raciti S 1989 Recent advances in analysis of laminated beams and plates, part I: Shear effects and buckling. *AIAA Journal*, **27**(7), 923–935.

Novozhilov VV 1961 *Theory of elasticity*. Pergamon.

Schardt R 1989 *Verallgemeinerte technische biegetheorie*. Springer.

Schardt R 1994 Generalized beam theory–an adequate method for coupled stability problems. *Thin-Walled Structures*, **19**, 161–180.

Silvestre N 2007 Generalised beam theory to analyse the buckling behaviour of circular cylindrical shells and tubes. *Thin-Walled Structures*, **45**(2), 185–198.

Washizu K 1968 *Variational methods in elasticity and plasticity*. Pergamon.

Further reading

4

EBBT, TBT, and CLEC in unified form

Classical EBBT and TBT, and fully linear CLEC, were presented in the previous chapters as singular cases with increasing accuracy of the kinematic models. The aim of the present chapter is to introduce the models mentioned above in a unified manner via a condensed notation that represents a basic step towards the Carrera Unified Formulation (CUF). First, CLEC will be rewritten adopting the new notation, then EBBT and TBT will be obtained as particular cases. Finally, a discussion on Poisson locking and its correction will be given.

4.1 Unified formulation of CLEC

A complete linear beam model was presented in the previous chapter and particular attention was given to the differences between CLEC and classical theories. The introduction of the CLEC finite element allowed us to introduce the stiffness matrix as in Equation 3.10. It is important to point out that the matrix can be considered as being composed of nine sub-matrices of dimension 3×3, as in

Beam Structures: Classical and Advanced Theories, First Edition. Erasmo Carrera, Gaetano Giunta and Marco Petrolo.

the following:

$$
\begin{array}{cccc}
 & 1 & x & z \\
1 & \begin{bmatrix} K_{xx}^{1,1} & K_{xy}^{1,1} & K_{xz}^{1,1} \\ K_{yx}^{1,1} & K_{yy}^{1,1} & K_{yz}^{1,1} \\ K_{zx}^{1,1} & K_{zy}^{1,1} & K_{zz}^{1,1} \end{bmatrix} & \cdots & \cdots \\
x & \cdots & \cdots & \begin{bmatrix} K_{xx}^{x,z} & K_{xy}^{x,z} & K_{xz}^{x,z} \\ K_{yx}^{x,z} & K_{yy}^{x,z} & K_{yz}^{x,z} \\ K_{zx}^{x,z} & K_{zy}^{x,z} & K_{zz}^{x,z} \end{bmatrix} \\
z & \cdots & \begin{bmatrix} K_{xx}^{z,x} & K_{xy}^{z,x} & K_{xz}^{z,x} \\ K_{yx}^{z,x} & K_{yy}^{z,x} & K_{yz}^{z,x} \\ K_{zx}^{z,x} & K_{zy}^{z,x} & K_{zz}^{z,x} \end{bmatrix} & \cdots
\end{array}
\tag{4.1}
$$

Each sub-matrix is characterized by a fixed couple of expansion functions that are used in the explicit computation of the integrals, as shown in Equation 3.11. It is extremely important to point out that the formal expression of each component of the sub-matrices does not depend on the expansion functions. That is, corresponding components of different sub-matrices have the same formal expression, as shown in the following:

$$
\begin{aligned}
K_{xx}^{1,1} &= C_{44} \int_{\Omega} 1 \cdot 1 \, d\Omega \int_{l} N_{i,_y} N_{j,_y} dy \\
K_{xx}^{x,z} &= C_{44} \int_{\Omega} x \cdot z \, d\Omega \int_{l} N_{i,_y} N_{j,_y} dy \\
K_{xx}^{1,z} &= C_{44} \int_{\Omega} 1 \cdot z \, d\Omega \int_{l} N_{i,_y} N_{j,_y} dy
\end{aligned}
\tag{4.2}
$$

This implies that the sub-matrix can be considered as a fundamental invariant nucleus which can be used to build the global stiffness matrix in an automatic way. Let us introduce the following notation for the expansion functions, F_τ:

$$
\begin{aligned}
F_{\tau=1} &= 1 \\
F_{\tau=2} &= x \\
F_{\tau=3} &= z
\end{aligned}
\tag{4.3}
$$

The displacement field in Equation 3.1 becomes

$$
\begin{aligned}
u_x &= F_1 \, u_{x_1} + F_2 \, u_{x_2} + F_3 \, u_{x_3} = F_\tau \, u_{x_\tau} \\
u_y &= F_1 \, u_{y_1} + F_2 \, u_{y_2} + F_3 \, u_{y_3} = F_\tau \, u_{y_\tau} \\
u_z &= F_1 \, u_{z_1} + F_2 \, u_{z_2} + F_3 \, u_{z_3} = F_\tau \, u_{z_\tau}
\end{aligned}
\tag{4.4}
$$

where the repeated indexes indicate summation according to the Einstein notation. The displacement vector can therefore be written as

$$\mathbf{u} = F_\tau \mathbf{u}_\tau, \quad \tau = 1, 2, 3 \tag{4.5}$$

If a FE formulation is introduced and two-node elements are adopted, the nodal unknown vector is given by

$$\mathbf{q}_{\tau i} = \left\{ q_{u_{x_{\tau i}}} \quad q_{u_{y_{\tau i}}} \quad q_{u_{z_{\tau i}}} \right\}^T \quad \tau = 1, 2, 3 \quad i = 1, 2 \tag{4.6}$$

The compact form of the internal work seen in Equation 3.9 becomes

$$\delta L_{int} = \delta \mathbf{q}_{\tau i}^T \mathbf{K}^{ij\tau s} \mathbf{q}_{sj} \tag{4.7}$$

where

- τ and s are the expansion function indexes;
- i and j are the shape function indexes.

Coherently with the introduced notation, the matrix in Equation 4.1 can, for a given i, j pair, be expressed as

$$\begin{array}{cccc}
 & s = 1 & s = 2 & s = 3 \\
\tau = 1 & \begin{bmatrix} K_{xx}^{11} & K_{xy}^{11} & K_{xz}^{11} \\ K_{yx}^{11} & K_{yy}^{11} & K_{yz}^{11} \\ K_{zx}^{11} & K_{zy}^{11} & K_{zz}^{11} \end{bmatrix} & \cdots & \cdots \\
\tau = 2 & \cdots & \cdots & \begin{bmatrix} K_{xx}^{23} & K_{xy}^{23} & K_{xz}^{23} \\ K_{yx}^{23} & K_{yy}^{23} & K_{yz}^{23} \\ K_{zx}^{23} & K_{zy}^{23} & K_{zz}^{23} \end{bmatrix} \\
\tau = 3 & \cdots & \begin{bmatrix} K_{xx}^{31} & K_{xy}^{31} & K_{xz}^{31} \\ K_{yx}^{31} & K_{yy}^{31} & K_{yz}^{31} \\ K_{zx}^{31} & K_{zy}^{31} & K_{zz}^{31} \end{bmatrix} & \cdots
\end{array} \tag{4.8}$$

Each 3×3 block is called a fundamental nucleus of the stiffness matrix. A component of the nucleus is given here for different combinations of expansion functions

$$\begin{aligned}
K_{xx}^{ij11} &= C_{44} \int_\Omega F_1 \cdot F_1 \, d\Omega \int_l N_{i,y} N_{j,y} dy \\
K_{xx}^{ij23} &= C_{44} \int_\Omega F_2 \cdot F_3 \, d\Omega \int_l N_{i,y} N_{j,y} dy \\
K_{xx}^{ij13} &= C_{44} \int_\Omega F_1 \cdot F_3 \, d\Omega \int_l N_{i,y} N_{j,y} dy
\end{aligned} \tag{4.9}$$

The unified formulation introduced is of particular interest when a computational implementation is considered. The exploitation of the four indexes in the formal expression of the fundamental nucleus makes it possible to compute the stiffness matrix by means of four nested *FOR*-cycles, that is, one for each index.

4.2 EBBT and TBT as particular cases of CLEC

EBBT and TBT are particular cases of CLEC, and they can therefore be obtained by acting on the full linear expansion. As far as the TBT is concerned, the displacement field is given by

$$
\begin{aligned}
u_x &= u_{x_1} \\
u_y &= u_{y_1} + x\, u_{y_2} + z\, u_{y_3} \\
u_z &= u_{z_1}
\end{aligned}
\tag{4.10}
$$

That is, a linear out-of-plane warping distribution is considered and constant in-plane displacement distributions are accounted for. Starting from the CLEC case, two possible techniques can be used to obtain TBT; (1) the rearranging of rows and columns of the stiffness matrix; (2) penalization of the stiffness terms related to u_{x_2}, u_{x_3}, u_{z_2}, and u_{z_3} (the latter is preferred here in the numerical applications). The main diagonal terms have to be considered, that is, $i = j$ and $\tau = s$; moreover, only the component with $\tau, s = 2, 3$ has to be penalized, therefore

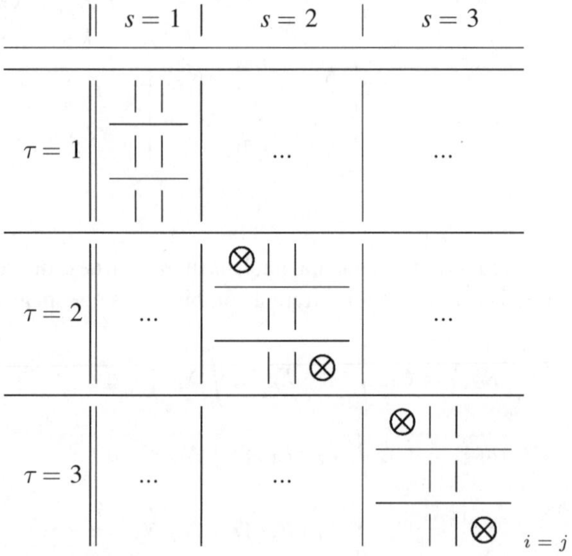

$$
\begin{array}{c||c|c|c}
 & s = 1 & s = 2 & s = 3 \\
\hline\hline
\tau = 1 & \begin{array}{c|c|c} & & \\ \hline & & \\ \hline & & \end{array} & \cdots & \cdots \\
\hline
\tau = 2 & \cdots & \begin{array}{c|c|c} \otimes & & \\ \hline & & \\ \hline & & \otimes \end{array} & \cdots \\
\hline
\tau = 3 & \cdots & \cdots & \begin{array}{c|c|c} \otimes & & \\ \hline & & \\ \hline & & \otimes \end{array}
\end{array}_{\,i = j}
\tag{4.11}
$$

The EBBT can be obtained through the penalization of ϵ_{xy} and ϵ_{zy}. This condition can be imposed by using a penalty value χ in the following constitutive equations:

$$\begin{aligned}\sigma_{xy} &= \chi C_{55}\epsilon_{xy} + \chi C_{45}\epsilon_{zy} \\ \sigma_{zy} &= \chi C_{45}\epsilon_{xy} + \chi C_{44}\epsilon_{zy}\end{aligned} \tag{4.12}$$

Each stiffness matrix component that contains C_{44}, C_{45}, or C_{55} will be penalized.

Example 4.2.1 *A simply supported square beam is considered and two length-to-thickness ratios, L/h, are considered:* 100 *and* 10. *Thin and moderately thick structures are therefore considered. The cross-section edge dimension is* 0.1 m. *Young's modulus, E, is equal to* 75 GPa. *The Poisson ratio, ν, is equal to* 0.33. *A concentrated load, P_z, is applied at the mid-span center point, and is equal to* 50 N. *A benchmark solution is obtained by means of the EBBT:*

$$u_{z_b} = \frac{1}{48}\frac{P_z L^3}{EI} \tag{4.13}$$

where I indicates the moment of inertia of the cross-section. The analysis is conducted by using 40 *four-node (B4), elements. The results in Table 4.1 show that:*

- *a good match is found with the reference solution;*
- *the adoption of TBT and CLEC provides significant advantages on the accuracy when a moderately thick beam is considered, whereas the EBBT is sufficient when a slender beam is analyzed.*

Table 4.1 u_z at the loading point of a simply supported square beam.

Equation 4.13	EBBT	TBT	CLEC
	$L/h = 100, u_z \times 10^3$		
1.667	1.667	1.667	1.667
	$L/h = 10, u_z \times 10^6$		
1.667	1.667	1.712	1.712

This table presents the vertical displacement of the loading point of a compact square beam; results from EBBT, TBT, and CLEC are reported for different slenderness ratios.

4.3 Poisson locking and its correction

Poisson locking (PL), is a phenomenon that arises due to coupling among the normal deformations ϵ_{xx}, ϵ_{yy}, and ϵ_{zz} as stated by the Poisson coefficients:

$$\nu_{ij} = -\frac{\epsilon_{jj}}{\epsilon_{ii}} \quad i, j = x, y, z \tag{4.14}$$

PL affects the accuracy of EBBT, TBT, and CLEC models; it will be described in this section and possible remedies will be given.

4.3.1 Kinematic considerations of strains

Let us first consider strain distributions derived from the kinematic models considered so far:

$$\begin{aligned}
\epsilon_{xx}^{kin} &= u_{x,x} = u_{x_2} && \text{if} \quad \text{CLEC} \\
&= 0 && \text{if} \quad \text{EBBT or TBT} \\
\epsilon_{yy}^{kin} &= u_{y,y} = u_{y_1,y} + x\, u_{y_2,y} + z\, u_{y_3,y} && \text{if} \quad \text{CLEC or EBBT or TBT} \\
\epsilon_{zz}^{kin} &= u_{z,z} = u_{z_3} && \text{if} \quad \text{CLEC} \\
&= 0 && \text{if} \quad \text{EBBT or TBT}
\end{aligned} \tag{4.15}$$

It should be noted that:

- the out-of-plane axial strain, ϵ_{yy}^{kin}, is linear;
- the in-plane axial strains, ϵ_{xx}^{kin} and ϵ_{zz}^{kin}, are null in classical models;
- CLEC provides a constant geometrical distribution of in-plane axial strains;
- to avoid constant distributions of ϵ_{xx}^{kin} and ϵ_{zz}^{kin}, the order of the expansion above the cross-section must be greater than two, $N \geq 2$.

4.3.2 Physical considerations of strains

Physical considerations can be made, starting from the constitutive relations between the in-plane and out-of-plane strains

$$\epsilon_{xx}^{con}, \epsilon_{zz}^{con} \propto \nu \epsilon_{yy}^{kin} \tag{4.16}$$

This means that the order of magnitude of the in-plane and out-of-plane normal strains is the same, that is, since EBBT, TBT, and CLEC consider linear distributions of ϵ_{yy}^{kin}, then ϵ_{xx}^{con} and ϵ_{zz}^{con} will also be linear. It is clear that a contradiction exists between kinematics and constitutive laws: while the former do not account for in-plane strain distributions, $\epsilon_{xx}^{kin}, \epsilon_{zz}^{kin} = 0$, the latter provide linear distributions of ϵ_{xx}^{con} and ϵ_{zz}^{con}. This contradiction generates the PL that is responsible for the poor convergence rates of EBBT, TBT, and CLEC. Two possible remedies will now be described to contrast PL (Carrera and Brischetto, 2008a, 2008b), that is, to contrast ϵ_{xx}^{con} and ϵ_{zz}^{con}.

4.3.3 First remedy: use of higher-order kinematics

The most "natural" way of overcoming PL is based on the use of kinematics, considering at least second-order terms to describe u_x and u_z

$$\begin{aligned} u_x &= u_{x_1} + x\, u_{x_2} + z\, u_{x_3} + x^2\, u_{x_4} + \cdots \\ u_z &= u_{z_1} + x\, u_{z_2} + z\, u_{z_3} + x^2\, u_{z_4} + \cdots \end{aligned} \tag{4.17}$$

This kind of solution implies a considerable increase in the awkwardness of the problem equations, compared to the classical model ones. In the following chapters, we will present a tool that can be used to overcome this important limitation.

4.3.4 Second remedy: modification of elastic coefficients

It is clear that PL originates from constitutive laws which state the intrinsic coupling between in- and out-of-plane strain components. The second remedy illustrated here is based on a proper modification of such laws.

The 3D constitutive relations between stresses, $\boldsymbol{\sigma}$, and strains, $\boldsymbol{\epsilon}$, are given by Hooke's law

$$\{\boldsymbol{\sigma}\} = [\mathbf{C}]\{\boldsymbol{\epsilon}\} \tag{4.18}$$

For isotropic materials, its explicit form is

$$\begin{Bmatrix} \sigma_{xx} \\ \sigma_{yy} \\ \sigma_{zz} \\ \sigma_{yz} \\ \sigma_{xz} \\ \sigma_{xy} \end{Bmatrix} = \begin{bmatrix} C_{11} & C_{12} & C_{13} & 0 & 0 & 0 \\ C_{12} & C_{22} & C_{23} & 0 & 0 & 0 \\ C_{13} & C_{23} & C_{33} & 0 & 0 & 0 \\ 0 & 0 & 0 & C_{44} & 0 & 0 \\ 0 & 0 & 0 & 0 & C_{55} & 0 \\ 0 & 0 & 0 & 0 & 0 & C_{66} \end{bmatrix} \begin{Bmatrix} \varepsilon_{xx} \\ \varepsilon_{yy} \\ \varepsilon_{zz} \\ \varepsilon_{yz} \\ \varepsilon_{xz} \\ \varepsilon_{xy} \end{Bmatrix} \tag{4.19}$$

The elastic coefficients are related to the engineering constants (Poisson's ν, Young's E, and the shear G moduli) as follows:

$$\begin{aligned} C_{11} = C_{22} = C_{33} &= \frac{E\,(1-\nu)}{(1+\nu)(1-2\nu)} \\ C_{12} = C_{13} = C_{23} &= \frac{\nu E}{(1+\nu)(1-2\nu)} \\ C_{44} = C_{55} = C_{66} &= G \end{aligned} \tag{4.20}$$

These are the coefficients that have to be modified in order to prevent PL. Classical plate theories correct PL by imposing that the out-of-plane normal stress is zero. This hypothesis yields reduced material stiffness coefficients which have to be accounted for in Hooke's law for the in-plane stress and strain components. PL correction is obtained for the beam theory in the same manner: σ_{xx} and σ_{zz} are assumed to be zero in Hooke's law, that is

$$\begin{cases} \sigma_{xx} = C_{11}\epsilon_{xx} + C_{12}\epsilon_{yy} + C_{13}\epsilon_{zz} = 0 \\ \sigma_{zz} = C_{13}\epsilon_{xx} + C_{23}\epsilon_{yy} + C_{33}\epsilon_{zz} = 0 \end{cases} \tag{4.21}$$

This system has to be solved with respect to ϵ_{xx} and ϵ_{zz}

$$\begin{cases} \epsilon_{xx} = \dfrac{C_{13}\,C_{23} - C_{12}\,C_{33}}{C_{11}\,C_{33} - C_{13}^2}\,\epsilon_{yy} \\ \epsilon_{zz} = \dfrac{C_{13}\,C_{12} - C_{23}\,C_{11}}{C_{11}\,C_{33} - C_{13}^2}\,\epsilon_{yy} \end{cases} \tag{4.22}$$

These expressions have to be inserted into Equation 4.19 in the σ_{yy} row

$$\sigma_{yy} = \underbrace{\left(C_{22} - C_{12}\,\frac{C_{33}\,C_{12} - C_{13}\,C_{23}}{C_{11}\,C_{33} - C_{13}^2} - C_{23}\,\frac{C_{23}\,C_{11} - C_{13}\,C_{12}}{C_{11}\,C_{33} - C_{13}^2} \right)}_{C'_{22}} \epsilon_{yy} \tag{4.23}$$

C'_{22} is the reduced elastic coefficient that has to be used to contrast PL. The procedure described above is also valid in the case of orthotropic materials; in this case, not only will C_{22} be reduced, but so will the elastic coefficient that relates σ_{yy} to the in-plane shear strain ϵ_{xz}. In the particular case of isotropic materials, the reduced elastic coefficient is equal to the Young's modulus:

$$C'_{22} = E \tag{4.24}$$

Some final remarks are necessary to highlight the better role of the PL correction:

- The correction of the material coefficients does not have a consistent theoretical proof.

Table 4.2 Effect of the PL correction on u_z (Carrera *et al.* 2010a).

Model	PL corrected	PL non-corrected
$L/h = 100$, $u_z \times 10^2$ m, $u_{z_b} \times 10^2 = -1.333$ m		
EBBT	−1.333	−0.901
TBT	−1.333	−0.901
CLEC	−1.333	−0.901
$L/h = 10$, $u_z \times 10^5$ m, $u_{z_b} \times 10^5 = -1.333$ m		
EBBT	−1.333	−0.901
TBT	−1.343	−0.909
CLEC	−1.343	−0.909

This table shows the beneficial effect of the PL correction in the case of EBBT, TBT, and CLEC models.

- This means that the adoption of reduced material coefficients does not necessarily lead to the exact 3D solution, as shown in Carrera et al. (2011), where the correction of PL led to a model with less bending stiffness than the correct one.
- A consistent way of preventing PL consists of adopting higher-order models, as will be shown in subsequent chapters.
- The correction of the material coefficients becomes detrimental as nonlinear terms are present in the displacement field, see Example 4.3.2.

Example 4.3.1 *Let us consider a cantilevered square cross-section beam loaded by a force,* $F_z = -50$ N, *applied at the center point of the free-tip cross-section. The cross-section edge, h, is* 0.2 m *long, and two slenderness ratios,* L/h, *are considered:* 100 *and* 10. *The material is isotropic with* $E = 75$ GPa *and* $\nu = 0.33$. *The results are obtained via a FE model having a* 40 *four-node element (B4), mesh. A benchmark solution is obtained by means of the EBBT:*

$$u_{z_b} = \frac{1}{3}\frac{F_z L^3}{EI} \tag{4.25}$$

where I indicates the moment of inertia of the cross-section. Table 4.2 shows the displacement values at the loading point for the EBBT, TBT, and CLEC models. The second column indicates the results obtained with the PL correction activated, while the third column reports the results with the PL correction deactivated. It can be seen how the correction of PL enhances the flexibility of the structure and the convergence rate. Similar results are obtained also in the case of free-vibration analyses (Carrera et al., *2011, 2011), in which the correction of PL makes the natural bending frequencies decrease.*

Table 4.3 Effect of the PL correction on a bilinear beam model (Carrera and Petrolo, 2010).

Correction	Bilinear beam
$u_z \times 10^2$ m, $u_{z_b} \times 10^2 = -1.333$ m	
Activated	−1.866
Deactivated	−1.115

This table shows the vertical displacement of a cantilevered beam obtained through a bilinear beam model to highlight the detrimental effect of the PL correction.

Example 4.3.2 *The beam model adopted in Example 4.3.1 is considered when $L/h = 100$. A bilinear beam model is considered*

$$
\begin{aligned}
u_x &= u_{x_1} + x\, u_{x_2} + z\, u_{x_3} + xz\, u_{x_4} \\
u_y &= u_{y_1} + x\, u_{y_2} + z\, u_{y_3} + xz\, u_{y_4} \\
u_z &= u_{z_1} + x\, u_{z_2} + z\, u_{z_3} + xz\, u_{z_4}
\end{aligned}
\tag{4.26}
$$

In the following chapters it will be shown how such a beam model can easily be obtained by means of the present unified beam formulation. Table 4.3 shows the loading point displacement obtained via the model in Equation 4.26 for active and non-active PL correction cases. It is clear that:

- *the correction of the PL is absolutely detrimental when a nonlinear term is present;*
- *the bilinear model provides a better solution of CLEC in the deactivated case. We will see in the next chapters how more refined models than the bilinear model are able to detect the exact solution with no need for PL correction.*

References

Carrera E and Brischetto S 2008a Analysis of thickness locking in classical, refined and mixed multilayered plate theories. *Composite Structures*, **82**(4), 549–562.

Carrera E and Brischetto S 2008b Analysis of thickness locking in classical, refined and mixed theories for layered shells. *Composite Structures*, **85**(1), 83–90.

Carrera E and Giunta G 2010 Refined beam theories based on a unified formulation. *International Journal of Applied Mechanics*, **2**(1), 117–143.

Carrera E and Petrolo M 2010 Refined beam elements with only displacement variables and plate/shell capabilities. Submitted.

Carrera E and Petrolo M 2011 On the effectiveness of higher-order terms in refined beam theories. *Journal of Applied Mechanics*, **78**(2). DOI: 10.1115/1.4002207.

Carrera E, Giunta G, Nali P, and Petrolo M 2010a Refined beam elements with arbitrary cross-section geometries. *Computers & Structures*, **88**(5–6), 283–293. DOI: 10.1016/j.compstruc.2009.11.002.

Carrera E, Petrolo M, and Nali P 2011 Unified formulation applied to free vibrations finite element analysis of beams with arbitrary section. *Shock and Vibrations*, **18**(3), 485–502. DOI: 10.3233/SAV-2010-0528.

Carrera E, Petrolo M, and Varello A 2011. Advanced beam formulations for free vibration analysis of conventional and joined wings. *Journal of Aerospace Engineering*, In Press. DOI: 10.1061/(ASCE)AS.1943-5525.0000130.

5

Carrera Unified Formulation and refined beam theories

Classical beam theories are based on a fixed number of variables; this number is usually related to the particular problem that has to be analyzed. The bending of slender beams, for instance, is well described by the Euler–Bernoulli model, which has three unknowns, whereas, torsion or thin-walled beam analysis requires more sophisticated theories with a larger number of variables. This modeling approach is therefore problem dependent since a beam theory is ad hoc built to face a particular structural analysis and, therefore, there is no guarantee it can be extended to other cases. The problem dependency of a beam model limits its application field, and its extension to enhanced models is not straightforward.

This chapter describes the theoretical layout of a novel unified approach that overcomes the limits of classical modeling techniques. Displacement fields are in fact obtained in a unified manner, regardless of the order of the theory, which is considered as an input of the analysis. The overcoming step, from a basic to a higher-order model, is immediate and does not require any ad hoc implementations. The unified formulation will be presented and then exploited to derive the governing equations in both strong- and weak forms.

Beam Structures: Classical and Advanced Theories, First Edition. Erasmo Carrera, Gaetano Giunta and Marco Petrolo.

5.1 Unified formulation

The unified formulation of the beam cross-section displacement field is described by an expansion of generic functions, F_τ,

$$\mathbf{u} = F_\tau \mathbf{u}_\tau, \qquad \tau = 1, 2, \ldots, M \tag{5.1}$$

where F_τ are functions of the cross-section coordinates x and z, $\mathbf{u}_\tau$ is the displacement vector, and M stands for the number of terms of the expansion. According to the Einstein notation, the repeated subscript τ indicates summation. The choice of F_τ and M is arbitrary, that is, different base functions of any order can be taken into account to model the kinematic field of a beam above the cross-section. One possible choice is related to the use of Taylor-like polynomials consisting of the 2D base $x^i\, z^j$, where i and j are positive integers. Table 5.1 presents M and F_τ as functions of the order of the beam model, N. Each row shows the expansion terms of an N-order theory. If a second-order model, $N = 2$, is chosen, M will be equal to six and the explicit expression of the cross-section displacement field will be

$$\begin{aligned} u_x &= u_{x_1} + x\, u_{x_2} + z\, u_{x_3} + x^2\, u_{x_4} + xz\, u_{x_5} + z^2\, u_{x_6} \\ u_y &= u_{y_1} + x\, u_{y_2} + z\, u_{y_3} + x^2\, u_{y_4} + xz\, u_{y_5} + z^2\, u_{y_6} \\ u_z &= u_{z_1} + x\, u_{z_2} + z\, u_{z_3} + x^2\, u_{z_4} + xz\, u_{z_5} + z^2\, u_{z_6} \end{aligned} \tag{5.2}$$

The beam model given in Equation 5.2 has 18 displacement variables: 3 constant, 6 linear, and 9 parabolic. Such a model is referred to as *full* since all the terms of the expansion are used. The CUF offers the opportunity of dealing with *reduced* models by exploiting a lower number of variables; an example of a reduced model is given by

$$\begin{aligned} u_x &= u_{x_1} + x\, u_{x_2} + \qquad\quad + x^2\, u_{x_4} + xz\, u_{x_5} \\ u_y &= u_{y_1} + x\, u_{y_2} + z\, u_{y_3} + x^2\, u_{y_4} + \qquad\quad + z^2\, u_{y_6} \\ u_z &= u_{z_1} + \qquad\quad + z\, u_{z_3} + \qquad\quad + xz\, u_{z_5} + z^2\, u_{z_6} \end{aligned} \tag{5.3}$$

Table 5.1 Taylor-like polynomials.

N	M	F_τ
0	1	$F_1 = 1$
1	3	$F_2 = x \;\; F_3 = z$
2	6	$F_4 = x^2 \;\; F_5 = xz \;\; F_6 = z^2$
3	10	$F_7 = x^3 \;\; F_8 = x^2 z \;\; F_9 = xz^2 \;\; F_{10} = z^3$
$\vdots$	$\vdots$	$\vdots$
N	$(N+1)(N+2)/2$	$F_{(N^2+N+2)/2} = x^N \;\; \ldots \;\; F_{(N+1)(N+2)/2} = z^N$

This table presents the compact form of the Taylor-like polynomials.

The beam model given in Equation 5.3 has 13 displacement variables: 3 constant, 4 linear, and 6 parabolic. By acting on the linear model, $N = 1$, the EBBT and TBT models are obtained as particular cases.

5.2 Governing equations

The governing equations are derived by means of the principle of virtual displacements (PVD). Starting from the unified form of the displacement field in Equation 5.1, stiffness, mass, and loading arrays will be obtained here in terms of fundamental nuclei whose form is independent of the order of the model. The closed form solution is addressed first, then the finite element formulation is described.

5.2.1 Strong form of the governing equations

The strong form of the governing differential equations and the boundary conditions are obtained by means of PVD:

$$\delta L_i = \delta L_p + \delta L_l \tag{5.4}$$

L_i represents the strain energy. L_p and L_l stand for the work due to a surface loading, $\mathbf{p}^k$, and a line loading, $\mathbf{l}^k$, that act on a k sub-domain. δ stands for a virtual variation.

5.2.1.1 Variation of the strain energy

Stress, $\boldsymbol{\sigma}$, and strain, $\boldsymbol{\epsilon}$, components are grouped as follows:

$$\begin{aligned} \boldsymbol{\sigma}_p &= \left\{ \sigma_{zz} \quad \sigma_{xx} \quad \sigma_{zx} \right\}^T, \quad \boldsymbol{\epsilon}_p = \left\{ \epsilon_{zz} \quad \epsilon_{xx} \quad \epsilon_{zx} \right\}^T \\ \boldsymbol{\sigma}_n &= \left\{ \sigma_{zy} \quad \sigma_{xy} \quad \sigma_{yy} \right\}^T, \quad \boldsymbol{\epsilon}_n = \left\{ \epsilon_{zy} \quad \epsilon_{xy} \quad \epsilon_{yy} \right\}^T \end{aligned} \tag{5.5}$$

The subscript "n" stands for terms lying on the cross-section, while "p" stands for terms lying on planes which are orthogonal to Ω. Linear strain–displacement relations are then rewritten as

$$\begin{aligned} \boldsymbol{\epsilon}_p &= \boldsymbol{D}_p \mathbf{u} \\ \boldsymbol{\epsilon}_n &= \boldsymbol{D}_n \mathbf{u} = (\boldsymbol{D}_{n\Omega} + \boldsymbol{D}_{ny}) \mathbf{u} \end{aligned} \tag{5.6}$$

with

$$\boldsymbol{D}_p = \begin{bmatrix} 0 & 0 & \frac{\partial}{\partial z} \\ \frac{\partial}{\partial x} & 0 & 0 \\ \frac{\partial}{\partial z} & 0 & \frac{\partial}{\partial x} \end{bmatrix}, \quad \boldsymbol{D}_{n\Omega} = \begin{bmatrix} 0 & \frac{\partial}{\partial z} & 0 \\ 0 & \frac{\partial}{\partial x} & 0 \\ 0 & 0 & 0 \end{bmatrix}, \quad \boldsymbol{D}_{ny} = \begin{bmatrix} 0 & 0 & \frac{\partial}{\partial y} \\ \frac{\partial}{\partial y} & 0 & 0 \\ 0 & \frac{\partial}{\partial y} & 0 \end{bmatrix} \tag{5.7}$$

Hooke's law becomes

$$\boldsymbol{\sigma} = \boldsymbol{C}\boldsymbol{\epsilon} \tag{5.8}$$

According to Equation 5.5, the previous equation becomes

$$\begin{aligned} \boldsymbol{\sigma}_p &= \tilde{\boldsymbol{C}}_{pp}\boldsymbol{\epsilon}_p + \tilde{\boldsymbol{C}}_{pn}\boldsymbol{\epsilon}_n \\ \boldsymbol{\sigma}_n &= \tilde{\boldsymbol{C}}_{np}\boldsymbol{\epsilon}_p + \tilde{\boldsymbol{C}}_{nn}\boldsymbol{\epsilon}_n \end{aligned} \tag{5.9}$$

In the case of isotropic materials the matrices $\tilde{\boldsymbol{C}}_{pp}$, $\tilde{\boldsymbol{C}}_{nn}$, $\tilde{\boldsymbol{C}}_{pn}$, and $\tilde{\boldsymbol{C}}_{np}$ are

$$\tilde{\boldsymbol{C}}_{pp} = \begin{bmatrix} \tilde{C}_{11} & \tilde{C}_{12} & 0 \\ \tilde{C}_{12} & \tilde{C}_{22} & 0 \\ 0 & 0 & \tilde{C}_{66} \end{bmatrix}, \quad \tilde{\boldsymbol{C}}_{nn} = \begin{bmatrix} \tilde{C}_{55} & 0 & 0 \\ 0 & \tilde{C}_{44} & 0 \\ 0 & 0 & \tilde{C}_{33} \end{bmatrix},$$

$$\tilde{\boldsymbol{C}}_{pn} = \tilde{\boldsymbol{C}}_{np}^T = \begin{bmatrix} 0 & 0 & \tilde{C}_{13} \\ 0 & 0 & \tilde{C}_{23} \\ 0 & 0 & 0 \end{bmatrix} \tag{5.10}$$

According to the grouping of the stress and strain components in Equation 5.5, the virtual variation of the strain energy is considered as the sum of two contributions:

$$\delta L_i = \int_l \int_\Omega \delta\boldsymbol{\epsilon}_n^T \boldsymbol{\sigma}_n \, d\Omega \, dy + \int_l \int_\Omega \delta\boldsymbol{\epsilon}_p^T \boldsymbol{\sigma}_p \, d\Omega \, dy \tag{5.11}$$

The virtual variation of the strain energy in compact vectorial form is

$$\delta L_i = \int_l \delta\mathbf{u}_\tau^T \, \mathbf{K}^{\tau s} \, \mathbf{u}_s \, dy + \delta\mathbf{u}_\tau^T \, \mathbf{\Pi}^{\tau s} \, \mathbf{u}_s \Big|_{y=0}^{y=L} \tag{5.12}$$

The components of the differential matrix $\mathbf{K}^{\tau s}$ are

$$\begin{aligned} K_{xx}^{\tau s} &= J_{\tau_{,x} s_{,x}}^{22k} + J_{\tau_{,z} s_{,z}}^{66k} - J_{\tau s}^{44k} \frac{\partial^2}{\partial y^2} \\ K_{yy}^{\tau s} &= J_{\tau_{,x} s_{,x}}^{44k} + J_{\tau_{,z} s_{,z}}^{55k} - J_{\tau s}^{33k} \frac{\partial^2}{\partial y^2} \\ K_{zz}^{\tau s} &= J_{\tau_{,z} s_{,z}}^{11k} + J_{\tau_{,x} s_{,x}}^{66k} - J_{\tau s}^{55k} \frac{\partial^2}{\partial y^2} \end{aligned} \tag{5.13}$$

$$K_{zx}^{\tau s} = J_{\tau_{,z}s_{,x}}^{12k} + J_{\tau_{,x}s_{,z}}^{66k} \quad K_{xz}^{\tau s} = J_{\tau_{,x}s_{,z}}^{12k} + J_{\tau_{,z}s_{,x}}^{66k}$$

$$K_{zy}^{\tau s} = \left(J_{\tau_{,z}s}^{13k} - J_{\tau s_{,z}}^{55k}\right)\frac{\partial}{\partial y} \quad K_{yz}^{\tau s} = -\left(J_{\tau s_{,z}}^{13k} - J_{\tau_{,z}s}^{55k}\right)\frac{\partial}{\partial y}$$

$$K_{xy}^{\tau s} = \left(J_{\tau_{,x}s}^{23k} - J_{\tau s_{,x}}^{44k}\right)\frac{\partial}{\partial y} \quad K_{yx}^{\tau s} = -\left(J_{\tau s_{,x}}^{23k} - J_{\tau_{,x}s}^{44k}\right)\frac{\partial}{\partial y}$$

The generic terms $J_{\tau_{,\phi}s_{,\xi}}^{ggk}$, $J_{\tau s}^{ggk}$, $J_{\tau_{,\phi}s}^{ghk}$, and $J_{\tau s_{,\phi}}^{ghk}$ are the cross-section inertial momenta of a k sub-domain:

$$J_{\tau_{,\phi}s_{,\xi}}^{ggk} = \int_{\Omega^k} C_{gg}^k F_{\tau_{,\phi}} F_{s_{,\xi}} \, d\Omega \quad J_{\tau s}^{ggk} = \int_{\Omega^k} C_{gg}^k F_\tau F_s \, d\Omega$$
$$J_{\tau_{,\phi}s}^{ghk} = \int_{\Omega^k} C_{gh}^k F_{\tau_{,\phi}} F_s \, d\Omega \quad J_{\tau s_{,\phi}}^{ghk} = \int_{\Omega^k} C_{gh}^k F_\tau F_{s_{,\phi}} \, d\Omega \qquad (5.14)$$

As far as the boundary conditions are concerned, the components of $\mathbf{\Pi}^{\tau s}$ are:

$$\Pi_{xx}^{\tau s} = J_{\tau s}^{44k}\frac{\partial}{\partial y}, \quad \Pi_{yy}^{\tau s} = J_{\tau s}^{33k}\frac{\partial}{\partial y}, \quad \Pi_{zz}^{\tau s} = J_{\tau s}^{55k}\frac{\partial}{\partial y}$$
$$\Pi_{zx}^{\tau s} = \Pi_{xz}^{\tau s} = 0 \qquad (5.15)$$
$$\Pi_{zy}^{\tau s} = J_{\tau s_{,z}}^{55k}, \quad \Pi_{xy}^{\tau s} = J_{\tau s_{,x}}^{44k}, \quad \Pi_{yz}^{\tau s} = J_{\tau s_{,z}}^{13k}, \quad \Pi_{yx}^{\tau s} = J_{\tau s_{,x}}^{23k}$$

5.2.1.2 Virtual work of the external loadings

The virtual work done by the external loadings is assumed to be due to a surface loading and a line loading. The components of a surface loading are

$$\mathbf{p}^{kT} = \left\{ p_{xx}^{k\pm} \quad p_{xy}^{k\pm} \quad p_{xz}^{k\pm} \quad p_{zx}^{k\pm} \quad p_{zy}^{k\pm} \quad p_{zz}^{k\pm} \right\} \qquad (5.16)$$

They act as shown in Figure 5.1. The lateral surfaces $\left\{S_\phi^{k\pm} : \phi = x, z\right\}$ of the beam are defined on the basis of the normal vector $\left\{n_\phi^{k\pm} : \phi = x, z\right\}$. A normal vector with the same orientation as the *x*- or *z*-axis identifies a positive lateral surface. The external virtual work due to $\mathbf{p}$ is

$$\delta L_p = \left(\delta L_{p_{xx}^\pm}^k + \delta L_{p_{xy}^\pm}^k + \delta L_{p_{xz}^\pm}^k + \delta L_{p_{zx}^\pm}^k + \delta L_{p_{zy}^\pm}^k + \delta L_{p_{zz}^\pm}^k\right)_k \qquad (5.17)$$

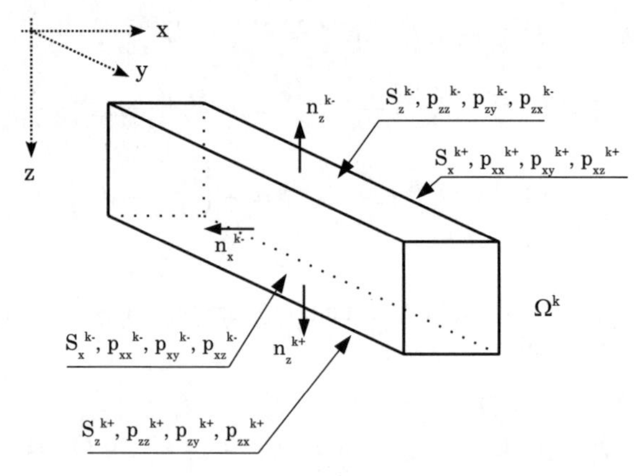

Figure 5.1 Components of a surface loading; lateral surfaces and normal vectors of the beam.

Its explicit terms are

$$\begin{aligned}
\left(\delta L^k_{p^\pm_{xx}},\ \delta L^k_{p^\pm_{zx}}\right) &= \int_l \delta u_{x\tau}\left(p^{k\pm}_{xx}E^{kz^\pm}_\tau,\ p^{k\pm}_{zx}E^{kx^\pm}_\tau\right)dy \\
\left(\delta L^k_{p^\pm_{zz}},\ \delta L^k_{p^\pm_{xz}}\right) &= \int_l \delta u_{z\tau}\left(p^{k\pm}_{zz}E^{kx^\pm}_\tau,\ p^{k\pm}_{xz}E^{kz^\pm}_\tau\right)dy \\
\left(\delta L^k_{p^\pm_{zy}},\ \delta L^k_{p^\pm_{xy}}\right) &= \int_l \delta u_{y\tau}\left(p^{k\pm}_{zy}E^{kx^\pm}_\tau,\ p^{k\pm}_{xy}E^{kz^\pm}_\tau\right)dy
\end{aligned} \tag{5.18}$$

where

$$\begin{aligned}
\left(E^{kx^+}_\tau, E^{kx^-}_\tau\right) &= \int_{x^k_1}^{x^k_2}\left(F_\tau\left(z^k_2, x\right), F_\tau\left(z^k_1, x\right)\right)dx \\
\left(E^{kz^+}_\tau, E^{kz^-}_\tau\right) &= \int_{z^k_1}^{z^k_2}\left(F_\tau\left(z, x^k_2\right), F_\tau\left(z, x^k_1\right)\right)dz
\end{aligned} \tag{5.19}$$

The components of a line loading (see Figure 5.2) are

$$\mathbf{l}^{kT} = \left\{ l^{k\pm}_{xx}\quad l^{k\pm}_{xy}\quad l^{k\pm}_{xz}\quad l^{k\pm}_{zx}\quad l^{k\pm}_{zy}\quad l^{k\pm}_{zz} \right\} \tag{5.20}$$

The external virtual work is

$$\delta L_l = \left(\delta L^k_{l^\pm_{xx}} + \delta L^k_{l^\pm_{zz}} + \delta L^k_{l^\pm_{zy}} + \delta L^k_{l^\pm_{xy}} + \delta L^k_{l^\pm_{zx}} + \delta L^k_{l^\pm_{xz}}\right)_k \tag{5.21}$$

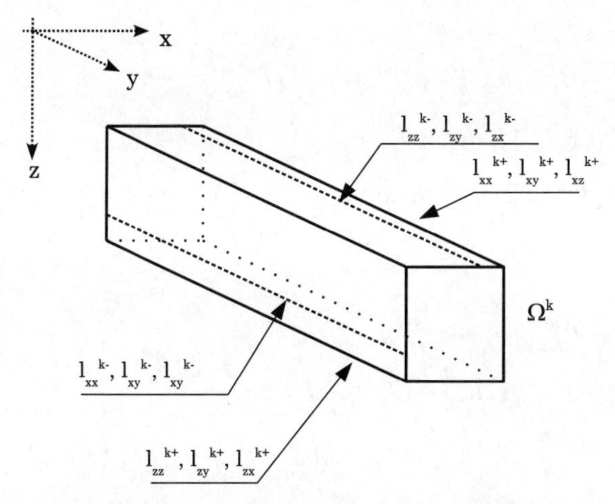

Figure 5.2 Components of a line loading.

whose terms are

$$\left(\delta L_{l_{zz}^{\pm}}^{k},\ \delta L_{l_{xz}^{\pm}}^{k}\right) = \int_{l} \delta u_{z\tau} \left(l_{zz}^{k\pm} F_{\tau}\left(z_{l_{zz}^{\pm}}^{k}, x_{l_{zz}^{\pm}}^{k}\right),\ l_{xz}^{k\pm} F_{\tau}\left(z_{l_{xz}^{\pm}}^{k}, x_{l_{xz}^{\pm}}^{k}\right)\right) dy$$

$$\left(\delta L_{l_{xx}^{\pm}}^{k},\ \delta L_{l_{zx}^{\pm}}^{k}\right) = \int_{l} \delta u_{x\tau} \left(l_{xx}^{k\pm} F_{\tau}\left(z_{l_{xx}^{\pm}}^{k}, x_{l_{xx}^{\pm}}^{k}\right),\ l_{zx}^{k\pm} F_{\tau}\left(z_{l_{zx}^{\pm}}^{k}, x_{l_{zx}^{\pm}}^{k}\right)\right) dy \tag{5.22}$$

$$\left(\delta L_{l_{zy}^{\pm}}^{k},\ \delta L_{l_{xy}^{\pm}}^{k}\right) = \int_{l} \delta u_{y\tau} \left(l_{zy}^{k\pm} F_{\tau}\left(z_{l_{zy}^{\pm}}^{k}, x_{l_{zy}^{\pm}}^{k}\right),\ l_{xy}^{k\pm} F_{\tau}\left(z_{l_{xy}^{\pm}}^{k}, x_{l_{xy}^{\pm}}^{k}\right)\right) dy$$

where $z_{l_{ij}^{\pm}}^{k}, x_{l_{ij}^{\pm}}^{k}$ are the coordinates of the line loading application point above a k cross-section sub-domain.

5.2.1.3 The fundamental nucleus of governing differential equations

The explicit form of the fundamental nucleus of the governing equations is obtained from Equations (5.13, 5.18, and 5.22):

$$\begin{aligned}
&\delta u_{x\tau}:\\
&\left(J_{\tau_{,x}s_{,z}}^{12k} + J_{\tau_{,z}s_{,x}}^{66k}\right) u_{zs} + \left(J_{\tau_{,x}s_{,x}}^{22k} + J_{\tau_{,z}s_{,z}}^{66k}\right) u_{xs}\\
&-J_{\tau s}^{44k} u_{xs_{,yy}} + \left(J_{\tau_{,x}s}^{23k} - J_{\tau s_{,x}}^{44k}\right) u_{ys_{,y}}\\
&= \left[p_{xx}^{k\pm} E_{\tau}^{kz^{\pm}} + p_{zx}^{k\pm} E_{\tau}^{kx^{\pm}} + l_{xx}^{k\pm} F_{\tau}\left(z_{l_{xx}^{\pm}}^{k}, x_{l_{xx}^{\pm}}^{k}\right) + l_{zx}^{k\pm} F_{\tau}\left(z_{l_{zx}^{\pm}}^{k}, x_{l_{zx}^{\pm}}^{k}\right)\right]_{k}
\end{aligned}$$

$$
\begin{aligned}
&\delta u_{y\tau}: \\
&-\left(J^{13k}_{\tau s,_z} - J^{55k}_{\tau,_z s}\right) u_{zs,_y} - \left(J^{13k}_{\tau s,_x} - J^{44k}_{\tau,_x s}\right) u_{xs,_y} \\
&+\left(J^{44k}_{\tau,_x s,_x} + J^{55k}_{\tau,_z s,_z}\right) u_{ys} - J^{33k}_{\tau s} u_{ys,_{yy}} \\
&= \left[p^{k\pm}_{zy} E^{kx^\pm}_{\tau} + p^{k\pm}_{xy} E^{kz^\pm}_{\tau} + l^{k\pm}_{zy} F_\tau\left(z^k_{l^\pm_{zy}}, x^k_{l^\pm_{zy}}\right) + l^{k\pm}_{xy} F_\tau\left(z^k_{l^\pm_{xy}}, x^k_{l^\pm_{xy}}\right)\right]_k
\end{aligned}
\tag{5.23}
$$

$$
\begin{aligned}
&\delta u_{z\tau}: \\
&\left(J^{11k}_{\tau,_z s,_z} + J^{66k}_{\tau,_x s,_x}\right) u_{zs} - J^{55k}_{\tau s} u_{zs,_{yy}} \\
&+\left(J^{12k}_{\tau,_z s,_x} + J^{66k}_{\tau,_x s,_z}\right) u_{xs} + \left(J^{13k}_{\tau,_z s} - J^{55k}_{\tau s,_z}\right) u_{ys,_y} \\
&= \left[p^{k\pm}_{zz} E^{kx^\pm}_{\tau} + p^{k\pm}_{xz} E^{kz^\pm}_{\tau} + l^{k\pm}_{zz} F_\tau\left(z^k_{l^\pm_{zz}}, x^k_{l^\pm_{zz}}\right) + l^{k\pm}_{xz} F_\tau\left(z^k_{l^\pm_{xz}}, x^k_{l^\pm_{xz}}\right)\right]_k
\end{aligned}
$$

The boundary conditions are

$$
\begin{aligned}
&\left[\delta u_{x\tau}\left(J^{44k}_{\tau s} u_{xs,_y} + J^{44k}_{\tau s,_x} u_{ys}\right)\right]^{y=L}_{y=0} = 0 \\
&\left[\delta u_{y\tau}\left(J^{13k}_{\tau s,_z} u_{zs} + J^{23k}_{\tau s,_x} u_{xs} + J^{33k}_{\tau s} u_{ys,_y}\right)\right]^{y=L}_{y=0} = 0 \\
&\left[\delta u_{z\tau}\left(J^{55k}_{\tau s} u_{zs,_y} + J^{55k}_{\tau s,_z} u_{ys}\right)\right]^{y=L}_{y=0} = 0
\end{aligned}
\tag{5.24}
$$

For a fixed approximation order, the nucleus has to be expanded versus the indexes τ and s in order to obtain the governing equations and the boundary conditions that concern the desired model. The assembly procedure for the stiffness matrix is based on the use of τ and s which are opportunely exploited to implement the FORTRAN statements. The core indexes are those related to the expansion functions F_τ and F_s; the stiffness matrix is computed by varying τ and s as shown in Figure 5.3.

5.2.1.4 Strong form analytical solution

The differential equations 5.23 and the related boundary conditions in Equations 5.24 are solved via a Navier-type solution by adopting the following displacement field:

$$
\begin{aligned}
u_{x\tau} &= U_{x\tau} F_\tau(x, z) \sin(\alpha y) \\
u_{y\tau} &= U_{y\tau} F_\tau(x, z) \sin(\alpha y) \\
u_{z\tau} &= U_{z\tau} F_\tau(x, z) \cos(\alpha y)
\end{aligned}
\tag{5.25}
$$

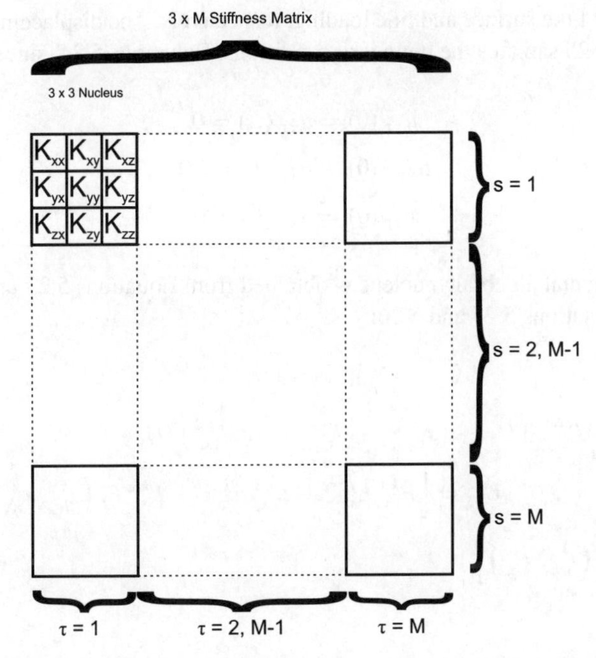

Figure 5.3 Graphic assembly procedure of the stiffness matrix.

We assume that the external loadings vary towards y in the following manner:

$$\mathbf{p}^k = \begin{Bmatrix} P_{xx}^{k\pm} \sin(\alpha y) \\ P_{xy}^{k\pm} \cos(\alpha y) \\ P_{xz}^{k\pm} \sin(\alpha y) \\ P_{zx}^{k\pm} \sin(\alpha y) \\ P_{zy}^{k\pm} \cos(\alpha y) \\ P_{zz}^{k\pm} \sin(\alpha y) \end{Bmatrix} \quad \mathbf{l}^k = \begin{Bmatrix} L_{xx}^{k\pm} \sin(\alpha y) \\ L_{xy}^{k\pm} \cos(\alpha y) \\ L_{xz}^{k\pm} \sin(\alpha y) \\ L_{zx}^{k\pm} \sin(\alpha y) \\ L_{zy}^{k\pm} \cos(\alpha y) \\ L_{zz}^{k\pm} \sin(\alpha y) \end{Bmatrix} \tag{5.26}$$

This last assumption does not represent a loss in generality, since a generic loading can be approximated via its Fourier series expansion (see Carrera and Giunta 2007, 2008). Term α is

$$\alpha = \frac{m\pi}{L} \tag{5.27}$$

where m represents the half-wave number along the beam axis. $U_{i\tau}$ are the maximal amplitudes of the displacement components and $P_{i,j}^{k\pm}$ and $L_{i,j}^{k\pm}$ the maximal amplitudes of the surface and line loading, respectively. The displacement field in Equations 5.25 satisfies the boundary conditions, Equations 5.24, since

$$\begin{aligned} u_{x\tau}(0) &= u_{x\tau}(L) = 0 \\ u_{y\tau,_y}(0) &= u_{y\tau,_y}(L) = 0 \\ u_{z\tau}(0) &= u_{z\tau}(L) = 0 \end{aligned} \tag{5.28}$$

The fundamental algebraic nucleus is obtained from Equations 5.23 upon substitution of Equations 5.25 and 5.26:

$$\begin{aligned}
&\delta U_{x\tau}: \\
&\left(J^{12k}_{\tau,_x s,_z} + J^{66k}_{\tau,_z s,_x}\right) U_{zs} + \left(J^{22k}_{\tau,_x s,_x} + J^{66k}_{\tau,_z s,_z} + \alpha^2 J^{44k}_{\tau s}\right) U_{xs} \\
&-\alpha\left(J^{23k}_{\tau,_x s} - J^{44k}_{\tau s,_x}\right) U_{ys} = \Big[P^{k\pm}_{xx} E^{kz^\pm}_\tau + P^{k\pm}_{zx} E^{kx^\pm}_\tau + L^{k\pm}_{xx} F_\tau\left(z^k_{l^\pm_{xx}}, x^k_{l^\pm_{xx}}\right) \\
&+L^{k\pm}_{zx} F_\tau\left(z^k_{l^\pm_{zx}}, x^k_{l^\pm_{zx}}\right)\Big]_k \\
&\delta U_{y\tau}: \\
&-\alpha\left(J^{13k}_{\tau s,_z} - J^{55k}_{\tau,_z s}\right) U_{zs} - \alpha\left(J^{23k}_{\tau s,_x} - J^{44k}_{\tau,_x s}\right) U_{xs} \\
&+\left(J^{44k}_{\tau,_x s,_x} + J^{55k}_{\tau,_z s,_z} + \alpha^2 J^{33k}_{\tau s}\right) U_{ys} \\
&= \Big[P^{k\pm}_{zy} E^{kx^\pm}_\tau + P^{k\pm}_{xy} E^{kz^\pm}_\tau + L^{k\pm}_{zy} F_\tau\left(z^k_{l^\pm_{zy}}, x^k_{l^\pm_{zy}}\right) + L^{k\pm}_{xy} F_\tau\left(x^k_{l^\pm_{xy}}, y^k_{l^\pm_{zy}}\right)\Big]_k \\
&\delta U_{z\tau}: \\
&\left(J^{11k}_{\tau,_z s,_z} + J^{66k}_{\tau,_x s,_x} + \alpha^2 J^{55k}_{\tau s}\right) U_{zs} + \left(J^{12k}_{\tau,_z s,_x} + J^{66k}_{\tau,_x s,_z}\right) U_{xs} \\
&-\alpha\left(J^{13k}_{\tau,_z s} - J^{55k}_{\tau s,_z}\right) U_{ys} = \Big[P^{k\pm}_{zz} E^{kx^\pm}_\tau + P^{k\pm}_{xz} E^{kz^\pm}_\tau + L^{k\pm}_{zz} F_\tau\left(z^k_{l^\pm_{zz}}, x^k_{l^\pm_{zz}}\right) \\
&+L^{k\pm}_{xz} F_\tau\left(z^k_{l^\pm_{xz}}, x^k_{l^\pm_{xz}}\right)\Big]_k
\end{aligned} \tag{5.29}$$

For a fixed approximation order, the algebraic system has to be assembled according to the summation indexes τ and s. Its solution yields the maximal displacement amplitudes. The strains are retrieved by the geometric relations, Equations 5.6, and the stresses via the generalized Hooke's law, Equations 5.9.

5.2.2 Weak form of the governing equations

The weak form of the governing equations is obtained by means of the finite element method (FEM), which allows one to overcome the limits of analytical solutions in terms of geometry, loading, and boundary conditions. The derivation of the governing FE equations begins with the definition of the nodal displacement vector

$$\mathbf{q}_{\tau i} = \left\{ q_{u_{x_{\tau i}}} \quad q_{u_{y_{\tau i}}} \quad q_{u_{z_{\tau i}}} \right\}^T, \quad \tau = 1, 2, \ldots, M, \; i = 1, 2, \ldots, N_{EN} \tag{5.30}$$

where the subscript "i" indicates the element node and N_{EN} stands for the number of nodes per element. If a linear model is considered ($N = 1$, $M = 3$), and a two-node element is adopted, the element unknowns will be

$$\mathbf{q}_{\tau i} = \begin{Bmatrix} q_{u_{x_{11}}} & q_{u_{y_{11}}} & q_{u_{z_{11}}} & q_{u_{x_{21}}} & q_{u_{y_{21}}} & q_{u_{z_{21}}} & q_{u_{x_{31}}} & q_{u_{y_{31}}} & q_{u_{z_{31}}} \\ q_{u_{x_{12}}} & q_{u_{y_{12}}} & q_{u_{z_{12}}} & q_{u_{x_{22}}} & q_{u_{y_{22}}} & q_{u_{z_{22}}} & q_{u_{x_{32}}} & q_{u_{y_{32}}} & q_{u_{z_{32}}} \end{Bmatrix}^T \tag{5.31}$$

The displacement variables are interpolated along the axis of the beam by means of the shape functions, N_i:

$$\mathbf{u} = N_i F_\tau \mathbf{q}_{\tau i} \tag{5.32}$$

Beam elements with two (B2), three (B3), and four (B4), nodes are considered here, whose shape functions are

$$N_1 = \tfrac{1}{2}(1 - r), \;\; N_2 = \tfrac{1}{2}(1 + r), \;\; \begin{cases} r_1 = -1 \\ r_2 = +1 \end{cases}$$

$$N_1 = \tfrac{1}{2}r(r - 1), \;\; N_2 = \tfrac{1}{2}r(r + 1), \;\; N_3 = -(1 + r)(1 - r), \;\; \begin{cases} r_1 = -1 \\ r_2 = +1 \\ r_3 = 0 \end{cases}$$

$$\begin{aligned} &N_1 = -\tfrac{9}{16}(r + \tfrac{1}{3})(r - \tfrac{1}{3})(r - 1), \;\; N_2 = \tfrac{9}{16}(r + \tfrac{1}{3})(r - \tfrac{1}{3})(r + 1), \\ &N_3 = +\tfrac{27}{16}(r + 1)(r - \tfrac{1}{3})(r - 1), \;\; N_4 = -\tfrac{27}{16}(r + 1)(r + \tfrac{1}{3})(r - 1), \end{aligned} \;\; \begin{cases} r_1 = -1 \\ r_2 = +1 \\ r_3 = -\tfrac{1}{3} \\ r_4 = +\tfrac{1}{3} \end{cases} \tag{5.33}$$

where the natural coordinate, r, varies from -1 to $+1$ and r_i indicates the position of the node within the natural beam boundaries. The beam model order is given by the expansion on the cross-section, and the number of nodes per element is related to the approximation along the longitudinal axis. An N-order beam model is therefore a theory that exploits an N-order Taylor-like polynomial to describe the kinematics of the beam cross-section. The choice of the beam model, the beam element, and the mesh (i.e., the number of beam elements), determines the total number of degrees of freedom of the structural model

$$\text{DOFs} = \underbrace{3 \times M}_{\text{number of DOFs per node}} \times [(\underbrace{N_{NE}}_{\text{number of nodes per element}} - 1) \times \underbrace{N_{BE}}_{\text{total number of beam elements}} + 1] \tag{5.34}$$

5.2.2.1 Stiffness matrix

The first step in assembling FE arrays is represented by the use of a proper variational statement. The PVD is exploited here

$$\delta L_{int} = \int_V (\delta \boldsymbol{\epsilon}_p^T \boldsymbol{\sigma}_p + \delta \boldsymbol{\epsilon}_n^T \boldsymbol{\sigma}_n) dV = \delta L_{ext} \tag{5.35}$$

where L_{int} stands for the strain energy, L_{ext} is the work of the external loadings, and δ stands for the virtual variation. Equations 5.6, 5.9, and 5.32 allow one to obtain a compact form of the virtual variation of the strain energy

$$\delta L_{int} = \delta \mathbf{q}_{\tau i}^T \mathbf{K}^{ij\tau s} \mathbf{q}_{sj} \tag{5.36}$$

where $\mathbf{K}^{ij\tau s}$ is the stiffness matrix written in the form of the fundamental nuclei. The fundamental nucleus is a 3×3 array which is formally independent of the order of the beam model. The nine components are

$$\begin{aligned}
K_{xx}^{ij\tau s} &= \tilde{C}_{22} \int_\Omega F_{\tau,x} F_{s,x} d\Omega \int_l N_i N_j dy + \tilde{C}_{66} \int_\Omega F_{\tau,z} F_{s,z} d\Omega \int_l N_i N_j dy \\
&\quad + \tilde{C}_{44} \int_\Omega F_\tau F_s d\Omega \int_l N_{i,y} N_{j,y} dy \\
K_{xy}^{ij\tau s} &= \tilde{C}_{23} \int_\Omega F_{\tau,x} F_s d\Omega \int_l N_i N_{j,y} dy + \tilde{C}_{44} \int_\Omega F_\tau F_{s,x} d\Omega \int_l N_{i,y} N_j dy \\
K_{xz}^{ij\tau s} &= \tilde{C}_{12} \int_\Omega F_{\tau,x} F_{s,z} d\Omega \int_l N_i N_j dy + \tilde{C}_{66} \int_\Omega F_{\tau,z} F_{s,x} d\Omega \int_l N_i N_j dy \\
K_{yx}^{ij\tau s} &= \tilde{C}_{44} \int_\Omega F_{\tau,x} F_s d\Omega \int_l N_i N_{j,y} dy + \tilde{C}_{23} \int_\Omega F_\tau F_{s,x} d\Omega \int_l N_{i,y} N_j dy \\
K_{yy}^{ij\tau s} &= \tilde{C}_{55} \int_\Omega F_{\tau,z} F_{s,z} d\Omega \int_l N_i N_j dy + \tilde{C}_{44} \int_\Omega F_{\tau,x} F_{s,x} d\Omega \int_l N_i N_j dy \\
&\quad + \tilde{C}_{33} \int_\Omega F_\tau F_s d\Omega \int_l N_{i,y} N_{j,y} dy \qquad (5.37) \\
K_{yz}^{ij\tau s} &= \tilde{C}_{55} \int_\Omega F_{\tau,z} F_s d\Omega \int_l N_i N_{j,y} dy + \tilde{C}_{13} \int_\Omega F_\tau F_{s,z} d\Omega \int_l N_{i,y} N_j dy
\end{aligned}$$

$$K_{zx}^{ij\tau s} = \tilde{C}_{12} \int_{\Omega} F_{\tau,z} F_{s,x} d\Omega \int_{l} N_i N_j dy + \tilde{C}_{66} \int_{\Omega} F_{\tau,x} F_{s,z} d\Omega \int_{l} N_i N_j dy$$

$$K_{zy}^{ij\tau s} = \tilde{C}_{13} \int_{\Omega} F_{\tau,z} F_{s} d\Omega \int_{l} N_i N_{j,y} dy + \tilde{C}_{55} \int_{\Omega} F_{\tau} F_{s,z} d\Omega \int_{l} N_{i,y} N_j dy$$

$$K_{zz}^{ij\tau s} = \tilde{C}_{11} \int_{\Omega} F_{\tau,z} F_{s,z} d\Omega \int_{l} N_i N_j dy + \tilde{C}_{66} \int_{\Omega} F_{\tau,x} F_{s,x} d\Omega \int_{l} N_i N_j dy + \tilde{C}_{55} \int_{\Omega} F_{\tau} F_{s} d\Omega \int_{l} N_{i,y} N_{j,y} dy$$

The assembly procedure for the stiffness matrix is based on the use of the four indexes τ, s, i, and j which are opportunely exploited to implement the FORTRAN statements. The core indexes are those related to the expansion functions F_τ and F_s, and the fundamental nucleus is computed by varying τ and s, as shown in Figure 5.4 where the construction of the so-called τs-Block, which coincides with the node stiffness matrix, can be observed. Each τs-Block is then inserted into the element stiffness matrix, as shown in Figure 5.5. The element stiffness matrix is derived from the assembly of all the ij-Blocks, as shown in Figure 5.6. Any-order beam theory can be computed since the definition of the order acts on the τs-loop.

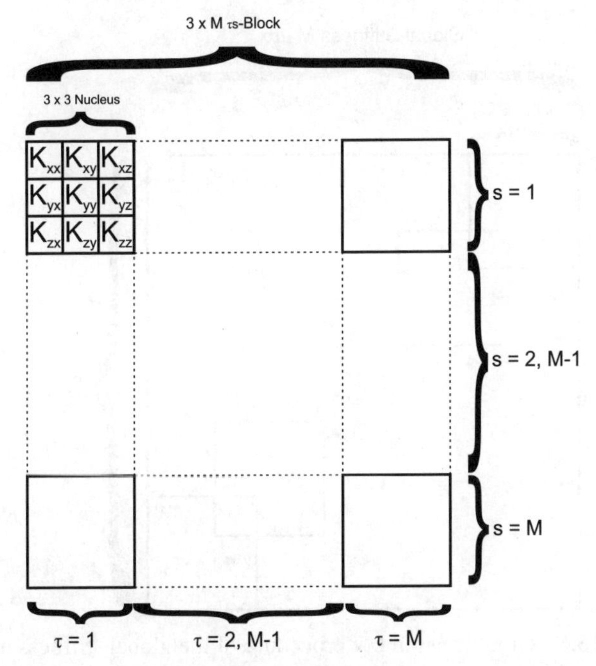

Figure 5.4 Graphic assembly procedure of the node stiffness matrix.

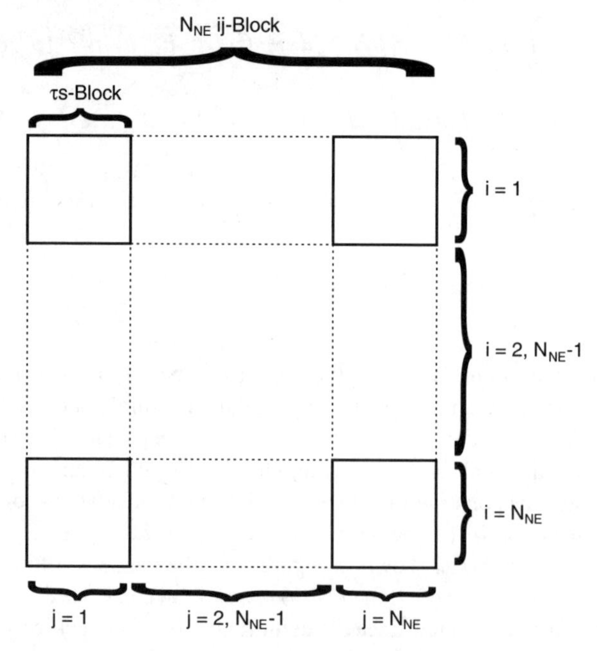

Figure 5.5 Graphic assembly procedure of the element stiffness matrix.

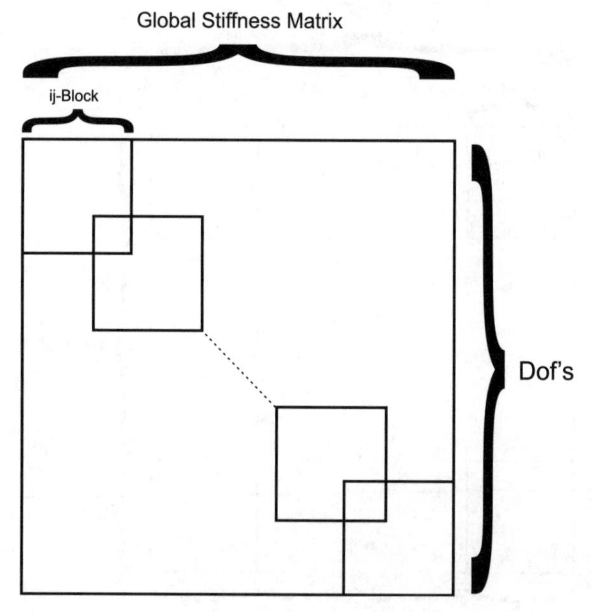

Figure 5.6 Graphic assembly procedure of the global stiffness matrix.

Example 5.2.1 *Let us consider a rectangular beam element, its length being L. A B2 element is used and a linear theory, $N = 1$, is adopted*

$$\tau, s = 1, 2, 3 \Rightarrow F_1 = 1, F_2 = x, F_3 = z$$
$$i, j = 1, 2 \Rightarrow N_1 = 1 - \frac{y}{L}, N_2 = \frac{y}{L}$$

The cross-section coordinates vary from $-a$ to $+a$ along the x-direction and from $-b$ to $+b$ along the z-direction. The K_{xx} component has to be computed for τ, $s = 2$, $i = 2$, and $j = 2$:

$$\begin{aligned} K_{xx}^{2122} &= \tilde{C}_{22} \int_{-a}^{+a} \int_{-b}^{+b} F_{2,_x} F_{2,_x} dxdz \int_0^L N_2 N_1 dy \\ &+ \tilde{C}_{66} \int_{-a}^{+a} \int_{-b}^{+b} F_{2,_z} F_{2,_z} dxdz \int_0^L N_2 N_1 dy \\ &+ \tilde{C}_{44} \int_{-a}^{+a} \int_{-b}^{+b} F_2 F_2 dxdz \int_0^L N_{2,_y} N_{1,_y} dy \end{aligned}$$

By substitution of the explicit expression of the functions the integrals become

$$\begin{aligned} K_{xx}^{2122} &= \tilde{C}_{22} \int_{-a}^{+a} \int_{-b}^{+b} 1 \cdot 1 \, dxdz \int_0^L \frac{y}{L}\left(1 - \frac{y}{L}\right) dy \\ &+ \tilde{C}_{66} \int_{-a}^{+a} \int_{-b}^{+b} 0 \cdot 0 \, dxdz \int_0^L \frac{y}{L}\left(1 - \frac{y}{L}\right) dy \\ &+ \tilde{C}_{44} \int_{-a}^{+a} \int_{-b}^{+b} x \cdot x \, dxdz \int_0^L \frac{1}{L}\left(-\frac{1}{L}\right) dy \end{aligned}$$

The final result is

$$K_{xx}^{2122} = \frac{2}{3} \tilde{C}_{22} \, a \, b \, L - \frac{4}{3} \tilde{C}_{44} \frac{b \, a^3}{L}$$

Figure 5.7 shows the position of the stiffness matrix element computed.

5.2.2.2 Mass matrix

The virtual variation of the work of the inertial loadings is

$$\delta L_{ine} = \int_V \rho \ddot{\boldsymbol{u}} \delta \boldsymbol{u}^T dV \tag{5.38}$$

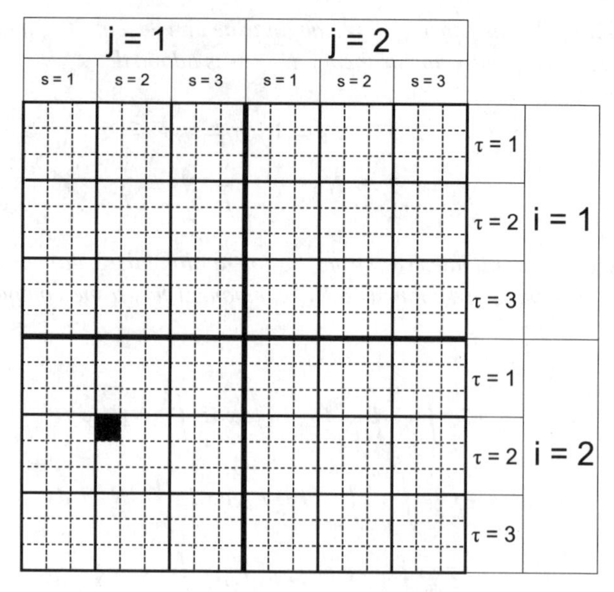

Figure 5.7 Position of K_{xx}^{2122} within the element stiffness matrix.

where ρ stands for the density of the material, and $\ddot{\boldsymbol{u}}$ is the acceleration vector. Equation 5.38 can now be rewritten using Equations 5.6 and 5.32:

$$\delta L_{ine} = \int_l \delta \mathbf{q}_{\tau i}^T N_i \left[\int_\Omega \rho (F_\tau \mathbf{I})(F_s \mathbf{I}) d\Omega \right] N_j \ddot{\boldsymbol{q}}_{sj} dy \tag{5.39}$$

where $\ddot{\boldsymbol{q}}$ is the nodal acceleration vector. The last equation can be rewritten in the following compact manner:

$$\delta L_{ine} = \delta \mathbf{q}_{\tau i}^T \mathbf{M}^{ij\tau s} \ddot{\boldsymbol{q}}_{sj} \tag{5.40}$$

where $\mathbf{M}^{ij\tau s}$ is the mass matrix in the form of the fundamental nucleus. Its components are

$$\begin{aligned} M_{xx}^{ij\tau s} &= M_{yy}^{ij\tau s} = M_{zz}^{ij\tau s} = \rho \int_\Omega F_\tau F_s d\Omega \int_l N_i N_j dy \\ M_{xy}^{ij\tau s} &= M_{xz}^{ij\tau s} = M_{yx}^{ij\tau s} = M_{yz}^{ij\tau s} = M_{zx}^{ij\tau s} = M_{zy}^{ij\tau s} = 0 \end{aligned} \tag{5.41}$$

The undamped dynamic problem can be written as

$$\boldsymbol{M}\ddot{\boldsymbol{a}} + \boldsymbol{K}\boldsymbol{a} = \boldsymbol{p} \tag{5.42}$$

where $\boldsymbol{a}$ is the vector of the nodal unknowns and $\boldsymbol{p}$ is the loading vector. Introducing harmonic solutions, it is possible to compute the natural frequencies, ω_i, for the homogeneous case, by solving an eigenvalue problem

$$(-\omega_i^2 \mathbf{M} + \mathbf{K})\boldsymbol{a}_i = 0 \tag{5.43}$$

where $\boldsymbol{a}_i$ is the ith eigenvector.

5.2.2.3 Loading vector

The application of surface, line, and point loads is discussed here to derive the equivalent loading vector. A generic surface load acting on a lateral face of the beam is considered first, $p_{\alpha\beta}(y)$, where α can be equal to x or z and β can be equal to x, y, or z. The first subscript, α, indicates the axis perpendicular to the surface, S_α, where the load is applied, whereas the second one, β, indicates the direction of the load. The virtual variation of the external work due to $p_{\alpha\beta}$ is given by

$$\delta L_{ext}^{p_{\alpha\beta}} = \int_{S_\alpha} \delta u_{\beta\tau}\, p_{\alpha\beta}\, d\alpha\, dy \tag{5.44}$$

By introducing the F_τ expansions and the nodal displacements, we obtain

$$\delta L_{ext}^{p_{\alpha\beta}} = \int_{S_\alpha} F_\tau(\alpha_p)\, N_i\, \delta q_{\beta_{\tau i}}\, p_{\alpha\beta}\, d\alpha\, dy \tag{5.45}$$

where α_p stands for the loading application coordinate.

A generic line load, $l_{\alpha\beta}(y)$, can be treated similarly. The virtual variation of the external work due to $l_{\alpha\beta}(y)$ is given by

$$\delta L_{ext}^{l_{\alpha\beta}} = \int_l \delta u_{\beta\tau}\, l_{\alpha\beta}\, dy \tag{5.46}$$

By introducing the F_τ expansions and the nodal displacements, we obtain

$$\delta L_{ext}^{l_{\alpha\beta}} = \int_l F_\tau(\alpha_p \beta_p)\, N_i\, \delta q_{\beta_{\tau i}}\, l_{\alpha\beta}\, dy \tag{5.47}$$

where α_p, β_p stand for the loading application coordinates above the cross-section.

The loading vector that is variationally coherent to the model, in the case of a generic concentrated load $\mathbf{P}$, is

$$\mathbf{P} = \left\{ P_{u_x} \quad P_{u_y} \quad P_{u_z} \right\}^T \tag{5.48}$$

the virtual work due to **P** is

$$\delta L_{ext} = \mathbf{P}\delta\mathbf{u}^T \tag{5.49}$$

and the virtual variation of **u** in the framework of the CUF is

$$\delta L_{ext} = F_\tau \mathbf{P}\delta\mathbf{u}_\tau^T \tag{5.50}$$

By introducing the nodal displacements and the shape functions, the previous equation becomes

$$\delta L_{ext} = F_\tau N_i \mathbf{P}\delta\mathbf{q}_{\tau i}^T \tag{5.51}$$

This equation permits us to identify the components of the nucleus which have to be loaded; that is, it permits the proper assembly of the loading vector by detecting the displacement variables that have to be loaded.

Example 5.2.2 *Let us consider the same beam element as in Example 5.2.1. A point load* **P** *acts on node* 1 *along the x-direction only*

$$\boldsymbol{P} = \left\{ P_{u_x} \quad 0 \quad 0 \right\}^T \tag{5.52}$$

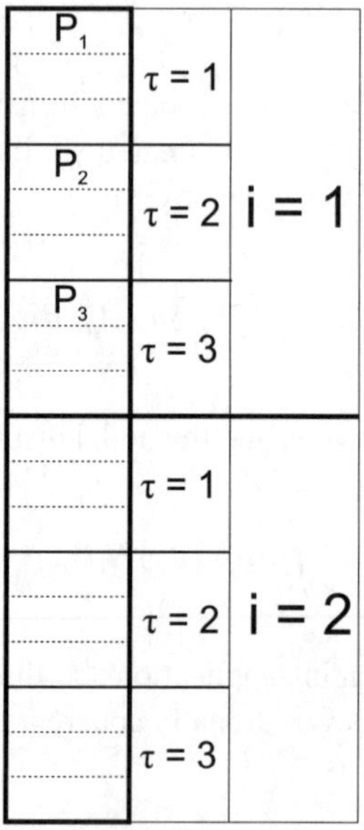

Figure 5.8 Position of the load components within the force vector.

The virtual variation of the external work is

$$\delta L_{ext} = \underbrace{P_{u_x}}_{P_1} \delta u_{x1} + \underbrace{x_p P_{u_x}}_{P_2} \delta u_{x2} + \underbrace{z_p P_{u_x}}_{P_3} \delta u_{x3} \tag{5.53}$$

where $[x_p, \; z_p]$ *are the coordinates on the cross-section of the loading application point. Figure 5.8 shows the position of the load components within the force vector.*

References

Carrera E and Giunta G. 2007 Hierarchical closed form solutions for plates bent by localized transverse loadings. *Journal of Zhejiang University SCIENCE A*, **8**(7), 1026–1037.

Carrera E and Giunta G. 2008 Hierarchical models for failure analysis of plates bent by distributed and localized transverse loadings. *Journal of Zhejiang University SCIENCE A*, **9**(5), 600–613.

Further reading

Carrera E and Nali P 2009 Mixed piezoelectric plate elements with direct evaluation of transverse electric displacement. *International Journal for Numerical Methods in Engineering*, **80**(4), 403–424.

Carrera E, Giunta G, Nali P, and Petrolo M 2010a Refined beam elements with arbitrary cross-section geometries. *Computers & Structures*, **88**(5–6), 283–293. DOI: 10.1016/j.compstruc.2009.11.002.

Carrera E, Giunta G, and Petrolo M 2010 A modern and compact way to formulate classical and advanced beam theories. *In: Developments in Computational Structures Technology*, Ch. 4. DOI: 10.4203/csets.25.4.

Carrera E, Petrolo M, and Nali P 2011 Unified formulation applied to free vibrations finite element analysis of beams with arbitrary section. *Shock and Vibrations*, **18**(3), 485–502. DOI: 10.3233/SAV-2010-0528.

Carrera E and Petrolo M 2011 On the effectiveness of higher-order terms in refined beam theories. *Journal of Applied Mechanics*, **78**(2). DOI: 10.1115/1.4002207.

Giunta G, Belouettar S, and Carrera E 2010 Analysis of FGM beams by means of classical and advanced theories. *Mechanics of Advanced Materials and Structures*, **17**(8), 622–635.

Giunta G, Biscani F, Carrera E, and Belouettar S 2011 Analysis of thin-walled beams via a one-dimensional unified formulation. *International Journal of Applied Mechanics*. In Press.

6

The parabolic, cubic, quartic, and N-order beam theories

The hierarchical capabilities of the present modern approach play a fundamental role in dealing with refined models in a compact unified manner. This leads us to the proper "tuning" of the order of the beam model with respect to the problem to be faced with no need for ad hoc formulations. While classical models are able to describe problems such as the bending of a compact cross-section beam, refined models are mandatory to describe the mechanical response in case of different loadings (e.g., torsion) or thin-walled structures.

A detailed description of various refined beam models is given in this chapter. First, the second-order model, $N = 2$, is carried out by describing the kinematic model. Then, the cubic, $N = 3$, and the quartic, $N = 4$, models will be described. Generic N-order models will be addressed finally.

6.1 The second-order beam model, $N = 2$

The second-order model exploits a parabolic expansion of the Taylor-like polynomials

$$
\begin{array}{ll|l|l}
u_x = & \underbrace{u_{x_1}}_{N=0} & + x u_{x_2} + z u_{x_3} & + x^2 u_{x_4} + xz u_{x_5} + z^2 u_{x_6} \\
u_y = & u_{y_1} & + x u_{y_2} + z u_{y_3} & + x^2 u_{y_4} + xz u_{y_5} + z^2 u_{y_6} \\
u_z = & u_{z_1} & + \underbrace{x u_{z_2} + z u_{z_3}}_{N=1} & + \underbrace{x^2 u_{z_4} + xz u_{z_5} + z^2 u_{z_6}}_{N=2}
\end{array}
\tag{6.1}
$$

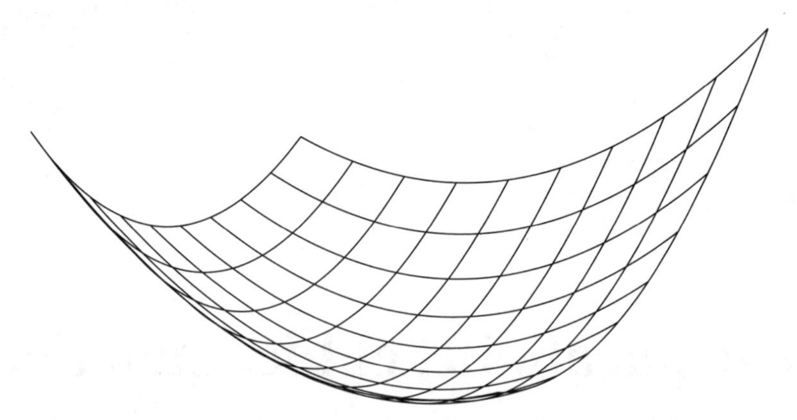

Figure 6.1 The parabolic distribution of a displacement component along the cross-section.

The beam model given by Equation (6.1) has 18 displacement variables: 3 constant ($N=0$), 6 linear ($N=1$), and 9 parabolic ($N=2$). Figure 6.1 shows the qualitative distribution of a displacement component above the cross-section. The strain components can be obtained straightforwardly; the normal components are

$$\begin{aligned}
\epsilon_{xx} &= \frac{\partial u_x}{\partial x} = u_{x_2} + 2\,x u_{x_4} + z u_{x_5} \\
\epsilon_{yy} &= \frac{\partial u_y}{\partial y} = u_{y_1,y} + x u_{y_2,y} + z u_{y_3,y} + x^2\, u_{y_4,y} + x z u_{y_5,y} + z^2\, u_{y_6,y} \\
\epsilon_{zz} &= \frac{\partial u_z}{\partial z} = u_{z_3} + z u_{z_5} + 2\,z u_{z_6}
\end{aligned} \tag{6.2}$$

The parabolic model leads to a linear distribution of ϵ_{xx} and ϵ_{zz} above the cross-section, and a second-order distribution of ϵ_{yy}. The distribution of ϵ_{xx} and ϵ_{zz} is important because it is related to the Poisson locking which is originated for $N \leq 1$. The same orders characterize the stress components which are computed via Hooke's law. The shear components can be similarly derived.

Example 6.1.1 *Let us consider the same B2 element as in Example 5.2.1 modeled via the $N=2$ theory. The normal strain components have to be computed at node 1 (i.e., $y_p = 0$) in a generic point of coordinates $[x_p, z_p]$. The choice of the beam theory allows us to derive the expansion functions*

$$\tau, s = 1, 2, 3, 4, 5, 6 \Rightarrow F_1 = 1,\, F_2 = x,\, F_3 = z,\, F_4 = x^2,\, F_5 = xz,\, F_6 = z$$

$$\tau, s = 1, 2, 3, 4, 5, 6 \Rightarrow F_{1,_x} = 0,\, F_{2,_x} = 1,\, F_{3,_x} = 0,\, F_{4,_x} = 2x,\, F_{5,_x} = z,\, F_{6,_x} = 0$$

$$\tau, s = 1, 2, 3, 4, 5, 6 \Rightarrow F_{1,_z} = 0,\, F_{2,_z} = 0,\, F_{3,_z} = 1,\, F_{4,_z} = 0,\, F_{5,_z} = x,\, F_{6,_z} = 2z$$

The use of a B2 element implies that

$$i, j = 1, 2 \Rightarrow N_1 = 1 - \frac{y}{L}, N_2 = \frac{y}{L}$$

$$i, j = 1, 2 \Rightarrow N_{1,_y} = -\frac{1}{L}, N_{2,_y} = \frac{1}{L}$$

The strain components are given by

$$\epsilon_{xx} = \frac{\partial u_x}{\partial x} = \frac{\partial (F_\tau N_i)}{\partial x} q_{x_{\tau i}} = F_{\tau,_x}(x_p, z_p) N_i(y_p) q_{x_{\tau i}}$$

$$\epsilon_{yy} = \frac{\partial u_y}{\partial y} = \frac{\partial (F_\tau N_i)}{\partial y} q_{y_{\tau i}} = F_\tau(x_p, z_p) N_{i,_y(y_p)} q_{y_{\tau i}}$$

$$\epsilon_{zz} = \frac{\partial u_z}{\partial z} = \frac{\partial (F_\tau N_i)}{\partial z} q_{z_{\tau i}} = F_{\tau,_z}(x_p, z_p) N_i(y_p) q_{z_{\tau i}}$$

Thus

$$\epsilon_{xx}(x_p, y_p, z_p) = 1\, q_{x_{21}} + 2\, x_p\, q_{x_{41}} + z_p\, q_{x_{51}}$$

$$\epsilon_{yy}(x_p, y_p, z_p) = -\frac{1}{L}(1\, q_{y_{11}} + x_p\, q_{y_{21}} + z_p\, q_{y_{31}} + x_p^2\, q_{y_{41}} + x_p\, z_p\, q_{y_{51}} + z_p^2\, q_{y_{61}})$$
$$+\frac{1}{L}(1\, q_{y_{12}} + x_p\, q_{y_{22}} + z_p\, q_{y_{32}} + x_p^2\, q_{y_{42}} + x_p\, z_p\, q_{y_{52}} + z_p^2\, q_{y_{62}})$$

$$\epsilon_{zz}(x_p, y_p, z_p) = 1\, q_{z_{31}} + x_p\, q_{z_{51}} + 2\, z_p\, q_{z_{61}}$$

6.2 The third-order, $N = 3$, and the fourth-order, $N = 4$, beam models

The third-order model exploits a cubic expansion of the Taylor-like polynomials

$$\begin{array}{ll} u_x = \ldots & \Bigg\| \; + x^3\, u_{x_7} + x^2\, z u_{x_8} + x z^2\, u_{x_9} + z^3\, u_{x_{10}} \\ u_y = \ldots & \Bigg\| \; + x^3\, u_{y_7} + x^2\, z u_{y_8} + x z^2\, u_{y_9} + z^3\, u_{y_{10}} \\ u_z = \underbrace{\ldots}_{N \le 2} & \Bigg\| \; + \underbrace{x^3\, u_{z_7} + x^2\, z u_{z_8} + x z^2\, u_{z_9} + z^3\, u_{z_{10}}}_{N=3} \end{array} \quad (6.3)$$

The beam model given by Equation (6.3) has 30 displacement variables: 12 cubic ($N = 3$) and 18 lower-order terms ($N \le 2$). Figure 6.2 shows the qualitative distribution of a displacement component above the cross-section. The strain

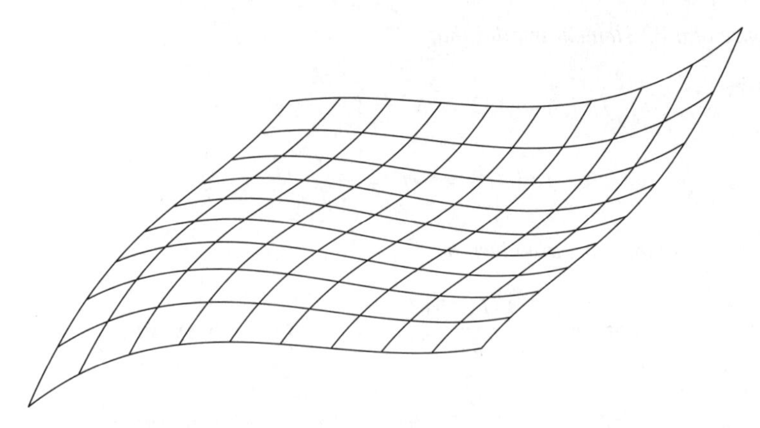

Figure 6.2 The cubic distribution of a displacement component along the cross-section.

components can be obtained straightforwardly; as an example the shear components are given:

$$
\begin{aligned}
\epsilon_{xy} &= \frac{\partial u_x}{\partial y} + \frac{\partial u_y}{\partial x} = \overbrace{\cdots}^{N\leq 2} \quad \Big\| \quad + x^3\, u_{x_{7},y} + x^2 z\, u_{x_{8},y} + x z^2\, u_{x_{9},y} + z^3\, u_{x_{10},y} \\
& \qquad\qquad\qquad\qquad\quad\;\; + 3\,x^2\, u_{y_7} + 2\,xz u_{y_8} + z^2\, u_{y_9} \\
\epsilon_{yz} &= \frac{\partial u_y}{\partial z} + \frac{\partial u_z}{\partial y} = \cdots \quad \Big\| \quad + x^2\, u_{y_8} + 2\,xz u_{y_9} + 3\,z^2\, u_{y_{10}} \\
& \qquad\qquad\qquad\qquad\quad\;\; + x^3\, u_{z_{7},y} + x^2 z u_{z_{8},y} + x z^2\, u_{z_{9},y} + z^3\, u_{z_{10},y} \\
\epsilon_{xz} &= \frac{\partial u_x}{\partial z} + \frac{\partial u_z}{\partial x} = \cdots \quad \Big\| \quad + x^2\, u_{x_8} + 2\,xz u_{x_9} + 3\,z^2\, u_{x_{10}} \\
& \qquad\qquad\qquad\qquad\quad\;\; + 3\,x^2\, u_{z_7} + 2\,xz u_{z_8} + z^2\, u_{z_9}
\end{aligned}
\tag{6.4}
$$

The fourth-order model exploits a quartic expansion of the Taylor-like polynomials

$$
\begin{aligned}
u_x &= \cdots \quad \Big\| \quad + x^4\, u_{x_{11}} + x^3\, z u_{x_{12}} + x^2\, z^2\, u_{x_{13}} + x z^3\, u_{x_{14}} + z^4\, u_{x_{15}} \\
u_y &= \cdots \quad \Big\| \quad + x^4\, u_{y_{11}} + x^3\, z u_{y_{12}} + x^2\, z^2\, u_{y_{13}} + x z^3\, u_{y_{14}} + z^4\, u_{y_{15}} \\
u_z &= \underbrace{\cdots}_{N\leq 3} \quad \Big\| \quad \underbrace{+ x^4\, u_{z_{11}} + x^3\, z u_{z_{12}} + x^2\, z^2\, u_{z_{13}} + x z^3\, u_{z_{14}} + z^4\, u_{z_{15}}}_{N=4}
\end{aligned}
\tag{6.5}
$$

The beam model given by Equation (6.5) has 45 displacement variables: 15 quartic ($N = 4$) and 30 lower-order terms ($N \leq 3$). Figure 6.3 shows the qualitative distribution of a displacement component above the cross-section. The strain

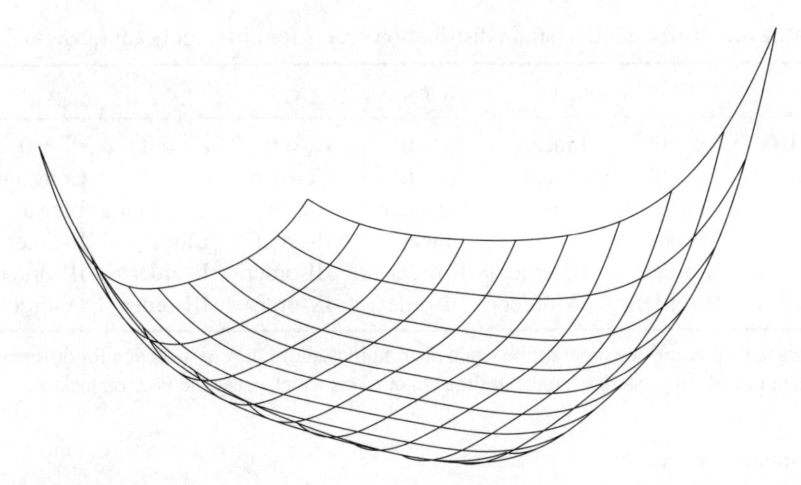

Figure 6.3 The $N = 4$ distribution of a displacement component along the cross-section.

components can be obtained straightforwardly; as an example the shear components are given:

$$
\begin{aligned}
\epsilon_{xy} = \frac{\partial u_x}{\partial y} + \frac{\partial u_y}{\partial x} = \overbrace{\cdots}^{N\leq 3} \quad & + x^4 u_{x_{11},y} + x^3 z u_{x_{12},y} + x^2 z^2 u_{x_{13},y} \\
& + x z^3 u_{x_{14},y} + z^4 u_{x_{15},y} + 4x^3 u_{y_{11}} + 3x^2 z u_{y_{12}} \\
& + 2xz^2 u_{y_{13}} + z^3 u_{y_{14}} \\
\epsilon_{yz} = \frac{\partial u_y}{\partial z} + \frac{\partial u_z}{\partial y} = \ldots \quad & + x^3 u_{y_{12}} + 2x^2 z u_{y_{13}} + 3xz^2 u_{y_{14}} + 3z^3 u_{y_{15}} \\
& + x^4 u_{z_{11},y} + x^3 z u_{z_{12},y} + x^2 z^2 u_{z_{13},y} \\
& + x z^3 u_{z_{14},y} + z^4 u_{z_{15},y} \\
\epsilon_{xz} = \frac{\partial u_x}{\partial z} + \frac{\partial u_z}{\partial x} = \ldots \quad & + x^3 u_{x_{12}} + 2x^2 z u_{x_{13}} + 3xz^2 u_{x_{14}} + 3z^3 u_{x_{15}} \\
& + 4x^3 u_{z_{11}} + 3x^2 z u_{z_{12}} + 2x z^2 u_{z_{13}} + z^3 u_{z_{14}}
\end{aligned}
\tag{6.6}
$$

Table 6.1 shows the strain distribution orders above the cross-section for various beam models. For $N \leq 1$ the Poisson locking is present and its correction is necessary.

6.3 *N*-order beam models

The present beam formulation is able to implement any-order theory by setting the order, N, as an input. An arbitrary refined model can be obtained by the following

Table 6.1 Cross-section strain distribution orders for different beam models.

	ϵ_{xx}	ϵ_{yy}	ϵ_{zz}	ϵ_{xy}	ϵ_{xz}	ϵ_{yz}
EBBT	$\epsilon_{xx} = 0^*$	Linear	$\epsilon_{zz} = 0^*$	$\epsilon_{xy} = 0$	$\epsilon_{xz} = 0$	$\epsilon_{yz} = 0$
TBT	$\epsilon_{xx} = 0^*$	Linear	$\epsilon_{zz} = 0^*$	Constant	$\epsilon_{xz} = 0$	Constant
$N = 1$	Constant*	Linear	Constant*	Linear	Constant	Linear
$N = 2$	Linear	II-order	Linear	II-order	Linear	II-order
$N = 3$	II-order	III-order	II-order	III-order	II-order	III-order
$N = 4$	III-order	IV-order	III-order	IV-order	III-order	IV-order

This table presents the order of the strain distributions above the cross-section for different beam models; the asterisk * indicates the strain terms which cause the Poisson locking.

compact expression:

$$\begin{aligned}
u_x &= \quad \cdots \quad \Bigg\| \quad + \sum_{M=0}^{N} x^{N-M} z^M u_{x\,\frac{N(N+1)}{2}+M+1} \\
u_y &= \quad \cdots \quad \Bigg\| \quad + \sum_{M=0}^{N} x^{N-M} z^M u_{y\,\frac{N(N+1)}{2}+M+1} \\
u_z &= \underbrace{\quad \cdots \quad}_{1,\ldots,N-1} \Bigg\| \quad \underbrace{+ \sum_{M=0}^{N} x^{N-M} z^M u_{z\,\frac{N(N+1)}{2}+M+1}}_{N\text{-order}}
\end{aligned} \tag{6.7}$$

Figure 6.4 shows the qualitative distribution of the seventh-order displacement field above the cross-section. The total number of displacement variables of the

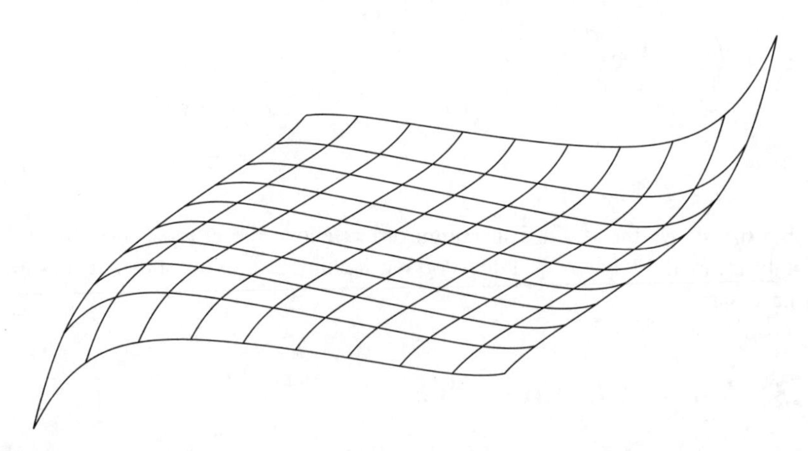

Figure 6.4 The $N = 7$ distribution of a displacement component along the cross-section.

model, N_{DV}, is related to N as

$$N_{DV} = 3 \times \frac{(N+1)(N+2)}{2} \tag{6.8}$$

In the case of the FE formulation, N_{DV} indicates the number of degrees of freedom per node. The strain components can be expressed in a compact manner as well

$$\begin{array}{ll||l}
\epsilon_{xx} = & \dots & + \sum_{M=0}^{N-1} (N-M)\, x^{N-M-1}\, z^M\, u_{x\,\frac{N(N+1)}{2}+M+1} \\
\epsilon_{yy} = & \dots & + \sum_{M=0}^{N} x^{N-M}\, z^M\, \frac{\partial \left(u_{y\,\frac{N(N+1)}{2}+M+1}\right)}{\partial y} \\
\epsilon_{zz} = & \dots & + \sum_{M=1}^{N} M\, x^{N-M}\, z^{M-1} u_{z\,\frac{N(N+1)}{2}+M+1} \\
\epsilon_{xy} = & \dots & + \sum_{M=0}^{N} x^{N-M}\, z^M\, \frac{\partial \left(u_{x\,\frac{N(N+1)}{2}+M+1}\right)}{\partial y} \\
 & & + \sum_{M=0}^{N-1} (N-M)\, x^{N-M-1}\, z^M\, u_{y\,\frac{N(N+1)}{2}+M+1} \\
\epsilon_{yz} = & \dots & + \sum_{M=0}^{N} x^{N-M}\, z^M\, \frac{\partial \left(u_{z\,\frac{N(N+1)}{2}+M+1}\right)}{\partial y} \\
 & & + \sum_{M=1}^{N} M\, x^{N-M}\, z^{M-1}\, u_{y\,\frac{N(N+1)}{2}+M+1} \\
\epsilon_{xz} = & \dots & + \sum_{M=0}^{N-1} (N-M)\, x^{N-M-1}\, z^M\, u_{z\,\frac{N(N+1)}{2}+M+1} \\
 & \underbrace{\dots}_{1,\dots,N-1} & \underbrace{+ \sum_{M=1}^{N} M\, x^{N-M}\, z^{M-1}\, u_{x\,\frac{N(N+1)}{2}+M+1}}_{N\text{-order}}
\end{array} \tag{6.9}$$

Further reading

Carrera E and Nali P 2009 Mixed piezoelectric plate elements with direct evaluation of transverse electric displacement. *International Joournal for Numerical Methods in Engineering*, **80**(4), 403–424.

Carrera E, Giunta G, Nali P, and Petrolo M 2010 Refined beam elements with arbitrary cross-section geometries. *Computers & Structures*, **88**(5–6), 283–293. DOI: 10.1016/j.compstruc.2009.11.002.

Carrera E, Giunta G, and Petrolo M 2010 A modern and compact way to formulate classical and advanced beam theories. In: *Developments in Computational Structures Technology*, Ch. 4. DOI: 10.4203/csets.25.4.

Carrera E, Petrolo M, and Nali P 2011 Unified formulation applied to free vibrations finite element analysis of beams with arbitrary section. *Shock and Vibrations*, **18**(3), 485–502. DOI: 10.3233/SAV-2010-0528.

Carrera E and Petrolo M 2011 On the effectiveness of higher-order terms in refined beam theories. *Journal of Applied Mechanics*, **78**(2), DOI: 10.1115/1.4002207.

Giunta G, Belouettar S, and Carrera E 2010 Analysis of FGM beams by means of a unified formulation. *IOP Conference Series: Material Science and Engineering*, **10**(1).

Giunta G, Belouettar S, and Carrera E 2010 Analysis of FGM beams by means of classical and advanced theories. *Mechanics of Advanced Materials and Structures*, **17**(8), 622–635.

Giunta G, Biscani F, Carrera E, and Belouettar S 2011 Analysis of thin-walled beams via a one-dimensional unified formulation. *International Journal of Applied Mechanics*. In Press.

7

CUF beam FE models: programming and implementation issue guidelines

CUF beam models are particularly suitable for the computer programming of FE codes. This is due to the hierarchical nature of the CUF matrices, which allows us to deal with arbitrary models in an automatic way. The implementation of an FEM code based on the CUF raises critical issues which have to be solved. Some of them are typical of every type of FE model, others are specific to the CUF, and the adopted solution represents one of the successful points of the entire formulation.

The main aim of this chapter is to offer an easy and complete guide to the most important programming and implementation issues related to the CUF in order to permit the reader to implement a CUF-based FE model on her/his own. The following main issues will be discussed:

1. preprocessing facts;
2. FE analysis;
3. postprocessing of results.

A set of numerical problems will be given in order to offer references for comparison purposes. Moreover, the given numerical examples will permit us to carry

Beam Structures: Classical and Advanced Theories, First Edition. Erasmo Carrera, Gaetano Giunta and Marco Petrolo.

out the analysis of critical issues such as convergence, numerical problems, and the accuracy of the results.

It should be mentioned that this chapter does not represent a detailed and comprehensive guide to implementing FE codes; its purpose is instead to describe details related to the CUF beam models adopted here to obtain the results shown in this book.

7.1 Preprocessing and input descriptions

Preprocessing indicates all those operations that are required to set up the input data necessary for a FE analysis. Attention is given here to all the data needed for the CUF FE model analysis, which can be divided into two groups: typical FE inputs and specific CUF inputs. The former indicate all those data that are commonly used in most FE codes. The latter indicate specific inputs needed for CUF models.

7.1.1 General FE inputs

7.1.1.1 Geometry and mesh data

The length of the beam, L, defines the geometry of the structure along the beam axis. As L is fixed, the structure has to be discretized, that is, a mesh has to be created. A mesh is usually defined according to:

- the type of beam element;
- the total number of beam elements, N_{BE}.

These two parameters are sufficient if homogeneous meshes are used; this means that constant length beam elements are adopted. The type of element parameter indicates the number of nodes per element, that is, the order of the shape functions. In the framework of the CUF, three beam elements are implemented:

- the two-node element, B2, which exploits linear shape functions;
- the three-node element, B3, which exploits parabolic shape functions;
- the four-node element, B4, which exploits cubic shape functions.

What the FE code receives as input is generally a set of two files: the node location file and the connectivity file. The first one contains the spanwise coordinates of each node, the second one defines the set of nodes of each beam element, Example 1 shows different mesh generations in CUF FEM models together with some comments concerning some critical issues. A convergence analysis is usually performed to define the mesh size of the problem.

Table 7.1 Node location list of the B2 mesh model.

Node ID	y-coordinate
1	0.0
2	0.5
3	1.0

This table presents the sample B2 mesh model node list related to Example 7.1.1.

Example 7.1.1 *Let us consider a beam with $L = 1$ m, three meshes have to be used composed of two beam elements the first mesh uses B2 elements, the second B3, and the third B4. In the case of B2 elements the total number of node is equal to*

$$N_{nodes} = (\underbrace{2}_{\text{number of nodes per element}} - 1) \times \underbrace{2}_{N_{BE}} + 1 = 3 \tag{7.1}$$

Table 7.1 shows the node location list. The first column indicates the identification number (ID) of each node, the second column shows the spanwise locations. Table 7.2 presents the connectivity of each B2 element. The first column shows the ID of the element, the other two columns define which of the nodes of the FE model is the first and second local node of an element. Figure 7.1 gives a graphic explanation of the difference between global and local nodes, that is, between the global node enumeration and the local one.

If two B3 elements are used the total number of nodes is given by

$$N_{nodes} = (\underbrace{3}_{\text{number of nodes per element}} - 1) \times \underbrace{2}_{N_{BE}} + 1 = 5 \tag{7.2}$$

Table 7.2 Connectivity of the B2 mesh model.

Element ID	Node 1	Node 2
1	1	2
2	2	3

This table presents the sample B2 mesh model connectivity related to Example 7.1.1.

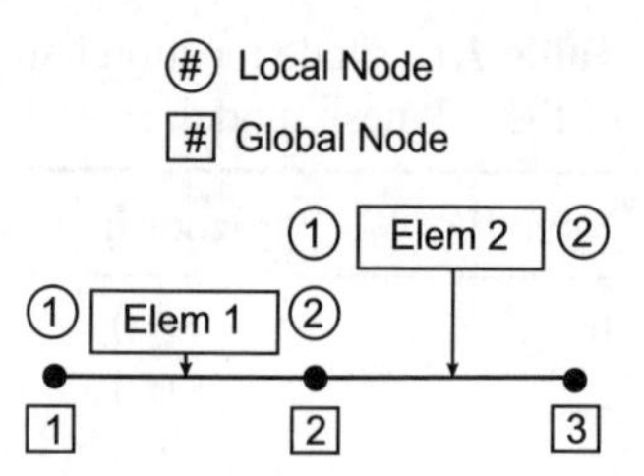

Figure 7.1 Global and local nodes for a B2 element.

Tables 7.3 and 7.4 report the node locations and the connectivity, respectively. If two B4 elements are used the total number of nodes is given by

$$N_{nodes} = (\underbrace{4}_{\text{number of nodes per element}} - 1) \times \underbrace{2}_{N_{BE}} + 1 = 7 \tag{7.3}$$

Tables 7.5 and 7.6 report the node locations and the connectivity, respectively. Figures 7.2 and 7.3 show the global and local nodes for B3 and B4 meshes, respectively. The order of the local nodes is related to the shape functions they are related to and, therefore, cannot be arbitrary.

Table 7.3 Node location list of the B3 mesh model.

Node ID	*y*-coordinate
1	0.0
2	0.5
3	1.0
4	0.25
5	0.75

This table presents the sample B3 mesh model node list related to Example 7.1.1.

Table 7.4 Connectivity of the B3 mesh model.

Element ID	Node 1	Node 2	Node 3
1	1	2	4
2	2	3	5

This table presents the sample B3 mesh model connectivity related to Example 7.1.1.

Table 7.5 Node location list of the B4 mesh model.

Node ID	y-coordinate
1	0.0
2	0.5
3	1.0
4	0.17
5	0.33
6	0.67
7	0.83

This table presents the sample B4 mesh model node list related to Example 7.1.1.

Table 7.6 Connectivity of the B4 mesh model.

Element ID	Node 1	Node 2	Node 3	Node 4
1	1	2	4	5
2	2	3	6	7

This table presents the sample B4 mesh model connectivity related to Example 7.1.1.

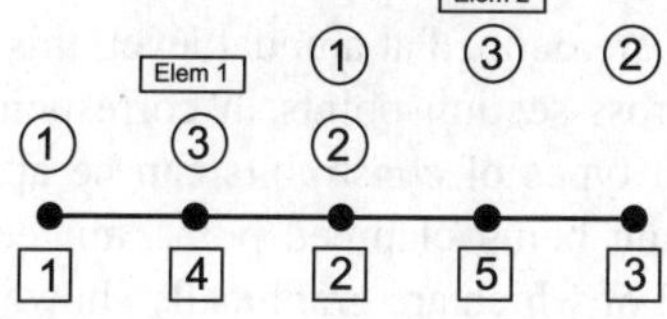

Figure 7.2 Global and local nodes for a B3 element.

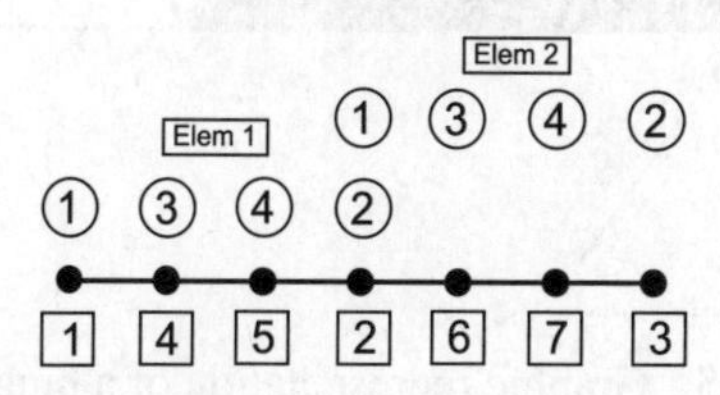

Figure 7.3 Global and local nodes for a B4 element.

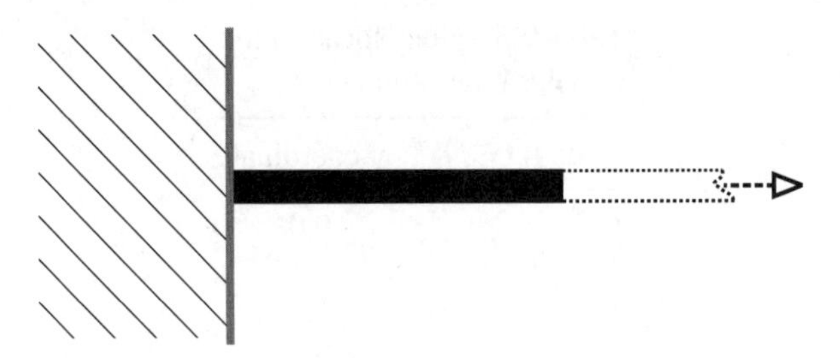

Figure 7.4 Graphic representation of a clamped node.

7.1.1.2 Loads and boundary conditions

A large variety of loading conditions can in general be used in the framework of a FE analysis, including mechanical, thermal, and inertial loads amongst others. A detailed analysis of loading models is not the aim of this book, but can be found in the excellent books by Bathe (1996) and Oñate (2009). The static FE analysis conducted in the examples given in this book uses point loads which, especially for thin-walled structures, represent severe test cases since they provoke a number of local effects that are generally hard to detect with 1D formulations. Only a few parameters are needed to define a point load:

- application point coordinates, $[x_p, y_p, z_p]$;
- load magnitude;
- load direction.

The procedure necessary to transfer the load to the FE model will be described in more detail in the next sections.

Boundary conditions are defined at a nodal level; this means that once a node is constrained, all the cross-section points in correspondence to the node will be constrained. Different types of constraints can be applied in the CUF beam model, the most important being clamped point, hinged, and hinged with free horizontal translation, all of which are graphically shown in Figures 7.4, 7.5, and 7.6, respectively. All the displacement components are locked in a clamped point; rotations are only allowed in a hinged point. If necessary, it is also possible to

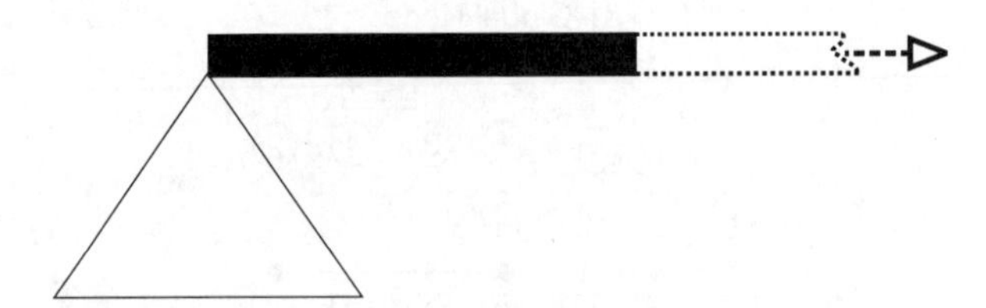

Figure 7.5 Graphic representation of a hinged node.

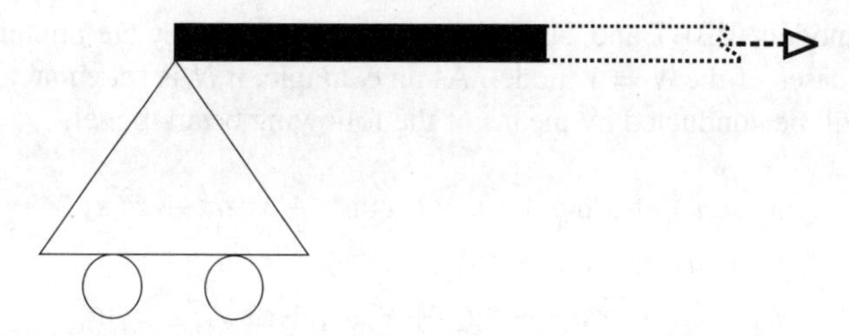

Figure 7.6 Graphic representation of a roll-hinged node.

constrain the given degrees of freedom of a certain point, as is usual in FE analysis. The constraining technique in the CUF will be discussed in the following sections.

7.1.1.3 Material properties

The definition of the material properties depends to a great extent on what kind of material is used, for example, isotropic, orthotropic, cross-ply laminates, etc. The present FE analyses are all conducted on isotropic made structures; this makes the definition of the material characteristics particularly easy since only Young's modulus, E, and the Poisson ratio, ν, are needed.

7.1.1.4 Type of analysis

This input defines the analysis that has to be conducted. In the present book two types of structural analysis are conducted: linear static, and linear free-vibration analysis. The definition of the analysis will automatically determine the FE matrices that have to be computed as well as the output files. In the case of linear static analysis the stiffness matrix and the loading force will be computed and as a result we obtain the displacement, strain, and stress fields. As far as the free-vibration analysis in concerned, stiffness and mass matrices will be computed for the eigenvalue analysis. Natural frequencies and modal shapes will be provided as output data.

7.1.2 Specific CUF inputs

This section describes the input data that are specifically needed for CUF FE models.

7.1.2.1 Order of the beam model

The possibility of freely choosing the order of the beam model is the most innovative feature of CUF models. The order, N, is set by the input and the analysis is then conducted by building all the FE matrices related to the chosen order.

Classical models, EBBT and TBT, can also be used and they are implemented as particular cases of the $N = 1$ model. As an example, if N is set equal to two, the analysis will be conducted by means of the following beam model:

$$
\begin{aligned}
u_x &= u_{x_1} + x\,u_{x_2} + z\,u_{x_3} + x^2\,u_{x_4} + xz\,u_{x_5} + z^2\,u_{x_6} \\
u_y &= u_{y_1} + x\,u_{y_2} + z\,u_{y_3} + x^2\,u_{y_4} + xz\,u_{y_5} + z^2\,u_{y_6} \\
u_z &= u_{z_1} + x\,u_{z_2} + z\,u_{z_3} + x^2\,u_{z_4} + xz\,u_{z_5} + z^2\,u_{z_6}
\end{aligned}
\tag{7.4}
$$

The CUF permits us to deal with reduced higher-order models since it is possible to choose which terms of a refined model have to be considered for the analysis as input. This means that the set of displacement variables that has to be retained can be inserted as input, thus making it possible to deal with beam models such as the following one:

$$
\begin{aligned}
u_x &= u_{x_1} + x\,u_{x_2} + x^2\,u_{x_4} + xz\,u_{x_5} \\
u_y &= z\,u_{y_3} + x^2\,u_{y_4} + xz\,u_{y_5} + z^2\,u_{y_6} \\
u_z &= u_{z_1} + x\,u_{z_2} + z\,u_{z_3} + z^2\,u_{z_6}
\end{aligned}
\tag{7.5}
$$

The choice of the beam model implies the definition of the model variables and the total number of unknowns of the problem, that is, the computational costs of the analysis. The proper choice of the order is, in general, problem dependent. A convergence analysis is usually conducted to establish which order of the beam model has to be used, as commonly done to define the mesh size of the problem.

Example 7.1.2 *Let us consider a cantilevered beam having a square cross-section of edge length equal to* 0.2 m *made of isotropic material with* $E = 75$ GPa *and* $\nu = 0.33$. *The structure is loaded with a vertical force equal to* -50 N. *A convergence analysis on the loading point vertical displacement regarding both mesh and beam order has to be performed for two values of the beam length L, namely,* 100 *and* 10; *that is, in the case of a slender and a moderately thick beam.*

Tables 7.7 and 7.8 show the results of the convergence study for the slender and the moderately thick beam, respectively. Rows are related to differently refined meshes while columns are related to increasing order beam models. Figure 7.7 shows the graphic convergence study for a fixed beam element type, B4, and different meshes and theories. Figure 7.8 shows the effect of the beam element type on the convergence. This convergence study highlights some fundamental and typical aspects related to the used of refined models:

- *The convergence towards the right solution moves from stiffer to more flexible models; this means that the refinement of the mesh as well as of the beam model leads to more flexible structures, that is, to larger deflections.*
- *Slender structures are well modeled by lower-order models, while thick beams need refined theories.*

Table 7.7 Doubly parametric convergence analysis for the slender beam (Carrera *et al.* 2010a).

No. of elements	EBBT	TBT	$N = 1$	$N = 2$
		$u_z \times 10^2$ m		
		B2		
1	−1.001	−1.001	−1.001	−0.893
3	−1.297	−1.297	−1.297	−1.165
5	−1.321	−1.321	−1.321	−1.236
10	−1.331	−1.331	−1.331	−1.287
40	−1.333	−1.333	−1.333	−1.323
		B3		
1	−1.333	−1.333	−1.333	−1.158
3	−1.333	−1.333	−1.333	−1.275
5	−1.333	−1.333	−1.333	−1.298
10	−1.333	−1.333	−1.333	−1.316
40	−1.333	−1.333	−1.333	−1.330
		B4		
1	−1.333	−1.333	−1.333	−1.239
3	−1.333	−1.333	−1.333	−1.302
5	−1.333	−1.333	−1.333	−1.315
10	−1.333	−1.333	−1.333	−1.325
40	−1.333	−1.333	−1.333	−1.332

This table shows the convergence analysis for the slender compact square beam by considering both the number of mesh elements and the order of the beam model.

- *Linear and classical models tend to have their own convergence behavior that is, in general, different from those of higher-order models. This is due to the Poisson locking correction that artificially enhances the flexibility of the beam.*
- *The Poisson locking correction can lead linear models to provide solutions that are different from those of higher-order models.*

In the case of free-vibration analyses, convergence studies can be performed on natural frequencies. Such studies lead to qualitatively similar results (Carrera et al. *2011).*

7.1.2.2 Cross-section geometry

The definition of the cross-section geometry is another specific point of CUF beam FE models. The present formulation is able to deal with arbitrary cross-section

Table 7.8 Doubly parametric convergence analysis for the moderately thick beam (Carrera *et al.* 2010a).

No of Elements	EBBT	TBT	$N = 1$	$N = 2$	$N = 3$	$N = 4$
			$u_z \times 10^5$ m			
			B2			
1	−1.001	−1.010	−1.010	−0.902	−0.904	−0.904
3	−1.297	−1.306	−1.306	−1.173	−1.176	−1.176
5	−1.321	−1.330	−1.330	−1.244	−1.246	−1.246
10	−1.331	−1.340	−1.340	−1.293	−1.296	−1.296
40	−1.333	−1.343	−1.343	−1.325	−1.327	−1.328
			B3			
1	−1.333	−1.343	−1.343	−1.166	−1.168	−1.168
3	−1.333	−1.343	−1.343	−1.283	−1.285	−1.285
5	−1.333	−1.343	−1.343	−1.305	−1.307	−1.307
10	−1.333	−1.343	−1.343	−1.321	−1.323	−1.324
40	−1.333	−1.343	−1.343	−1.329	−1.331	−1.333
			B4			
1	−1.333	−1.343	−1.343	−1.248	−1.250	−1.250
3	−1.333	−1.343	−1.343	−1.309	−1.311	−1.311
5	−1.333	−1.343	−1.343	−1.320	−1.322	−1.323
10	−1.333	−1.343	−1.343	−1.327	−1.329	−1.330
40	−1.333	−1.343	−1.343	−1.330	−1.332	−1.333

This table shows the convergence analysis for the moderately thick compact square beam by considering both the number of mesh elements and the order of the beam model.

geometries, that is, it can analyze beams with arbitrary complex shapes. The cross-section geometry is directly involved in the computation of the FE matrix elements where surface integrals have to be computed in the form of

$$\int_{\Omega} F_{\tau}(x, z) \cdot F_{s}(x, z)\, d\Omega$$

These integrals are computed numerically using a technique that will be described in the following sections. As far as the related input data are concerned, the numerical computation of the surface integral needs the cross-section to be discretized into a number of triangular elements. Such a discretization can be performed using a common FE preprocessor. The data needed in the present FE code are the coordinates of each node above the cross-section and the connectivity of the triangular elements. An example of a numerical mesh for a three-cell airfoil-shaped beam cross-section is shown in Figure 7.9. The number of triangular elements has to be

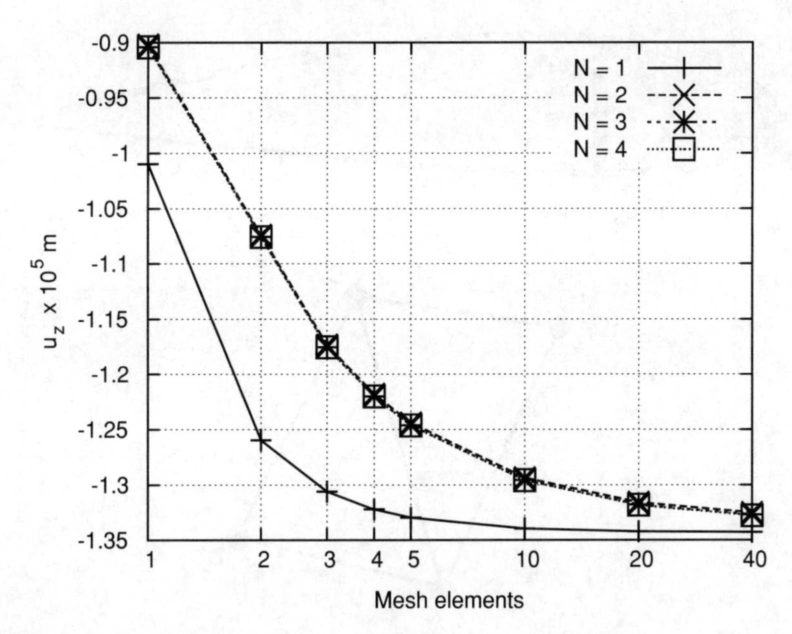

Figure 7.7 Convergence study for different beam models and meshes via B4 elements for the moderately thick beam (Carrera *et al*. 2010a).

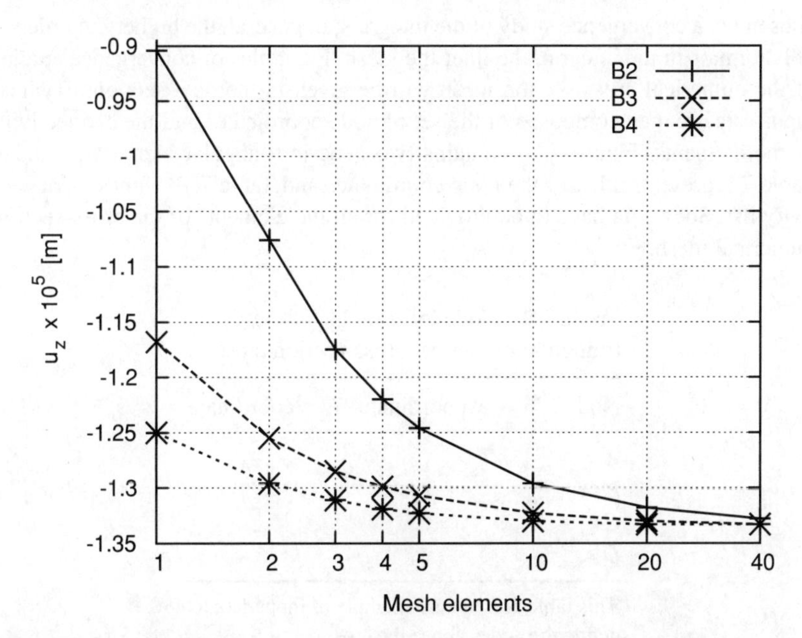

Figure 7.8 Convergence study for different meshes via a $N = 4$ model for the moderately thick beam (Carrera *et al*. 2010a).

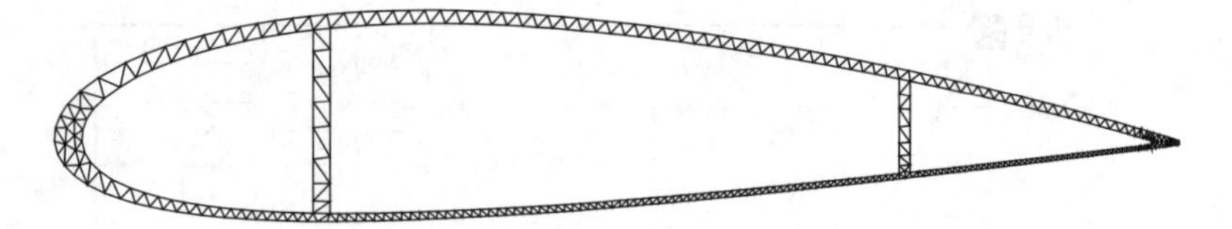

Figure 7.9 Triangular mesh for an airfoil-shaped cross-section having three cells.

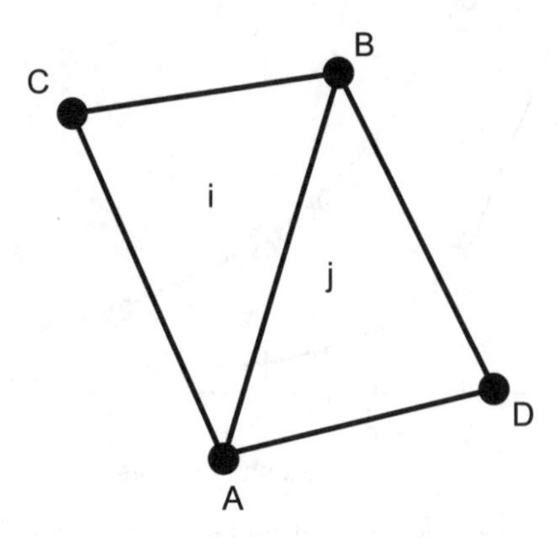

Figure 7.10 Two triangular elements of a cross-section numerical mesh.

chosen via a convergence study of the integrals; in general, the higher the order of polynomials in the integral, the finer the mesh. Examples of convergence studies of the numerical cross-section mesh will be given in the next sections. Typical input data are then composed of the set of node coordinates and the connectivity of the elements. Figure 7.10 contains two generic triangular elements, i and j. Table 7.9 presents a list of the node coordinates and Table 7.10 shows a connectivity list. Such data have to be provided for all the elements of the cross-section numerical mesh.

Table 7.9 Node location list of the triangular numerical cross-section mesh.

Node ID	x-coordinate	z-coordinate
A	x_A	z_A
B	x_B	z_B
C	x_C	z_C
D	x_D	z_D

This table presents an example of input data to define the nodes above the cross-section numerical mesh, see Figure 7.10 for a graphic description of the points.

Table 7.10 Connectivity of the triangular cross-section numerical mesh.

Element ID	Node 1	Node 2	Node 3
i	A	B	C
j	A	D	B

This table presents an example of connectivity input for the triangular elements of the cross-section numerical mesh as in Figure 7.10.

7.2 FEM code

This section gives guidelines that can be used to build a FE code based on CUF beam models. As in the previous sections of this chapter, attention is focused on particular issues related to the CUF implementation; the aim is not to give a comprehensive FE programming guide.

7.2.1 Stiffness and mass matrix

The stiffness and mass matrix implementation is described here by means of the nucleus-based structure. Critical issues related to the numerical computation of the integrals involved in the matrices are also discussed.

7.2.1.1 Nucleus-based implementation

The core base of the CUF beam model is represented by the hierarchical implementation of the FE matrices, which are expressed in terms of the so-called fundamental nuclei that are formally independent of the order of the beam model. The stiffness and mass matrix components are expressed as

$$K_{lk}^{ij\tau s}, \quad l, k = x, y, z$$
$$M_{lk}^{ij\tau s}, \quad l, k = x, y, z$$

It is clear that the computation of the nodal matrices has to be performed by means of the six indexes: i, j, τ, s, l, and k:

- i, j are the shape function indexes, and they therefore depend on the type of beam element adopted. They vary from one to the number of nodes per element (two for B2, three for B3, and four for B4).
- τ, s are the expansion function indexes, and they depend on the beam model order, N, adopted. They vary from one to $(N + 1)\times(N + 2)/2$.

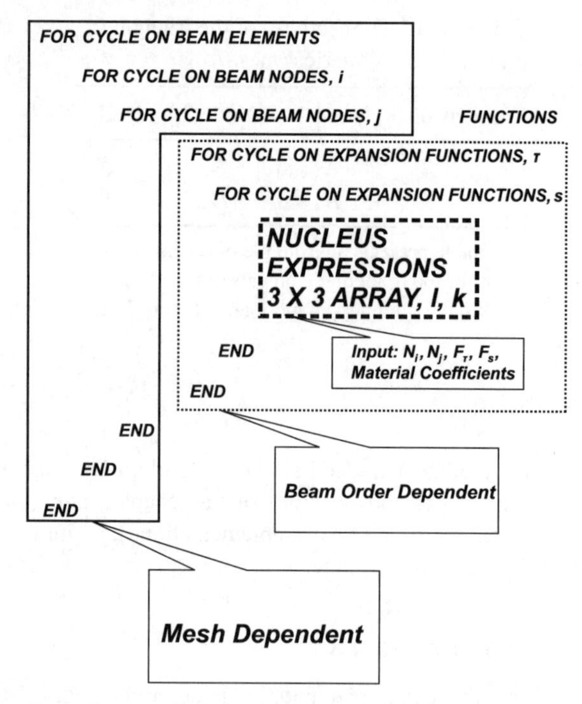

Figure 7.11 Graphic description of the hierarchical structure of the nucleus-based formulation.

- l, k are the indexes needed to span all the nine components of the nucleus, $l, k = x, y, z$. It should be pointed out that the number of components of the nucleus is related to the fact that the unknowns of the problems are the generalized displacement variables. If other primary unknowns are chosen, the dimension of the nucleus could in general change (Carrera and Nali, 2009).

A graphic description of the implementation of the hierarchical structure of CUF FE models is given in Figure 7.11. The outer cycles show the beam mesh characteristics, whereas the inner ones are related to the beam order features. The core of the procedure is represented by the computation of the nucleus components, which involves line and surface integrals

$$\int_l N_i N_j dy$$

$$\int_\Omega F_\tau(x, z) \cdot F_s(x, z)\, d\Omega$$

These integrals are computed numerically; the adopted computing techniques are described in the following sections.

7.2.1.2 Line integrals

Line integrals of shape functions are evaluated numerically by means of the Gauss quadrature technique. This method permits us to integrate polynomial functions precisely (Bathe 1996; Oñate, 2009) and is based on the following approximation:

$$\int_{-1}^{+1} f(\eta)\,d\eta \simeq \sum_{i=1}^{k} f(\eta_i)\,W_i \tag{7.6}$$

where k is the quadrature order, η_i are the Gauss points, and W_i are the weights. A quadrature of order k integrates a polynomial of order $(2k-1)$ precisely. Table 7.11 gives the Gauss points and weights for the first, second, and third quadrature orders.

Line integrals are computed over each element, that is, each integral domain is defined by the element boundaries. In order to be able to use the formula in Equation 7.6, the integration domain has to be $-1 \leq \eta \leq +1$, which is usually called the natural domain of the element. The Jacobian, J_E, of the element has to be calculated to compute the integral

$$\int_{l_E} f(y)\,dy = \int_{-1}^{+1} J_E(\eta)\,f(\eta)\,d\eta \simeq \sum_{i=1}^{k} J_E(\eta_i)\,f(\eta_i)\,W_i \tag{7.7}$$

Table 7.11 Gauss quadrature formula points and weights.

k	η_i	W_i
1	0.0	2.0
2	+0.577 350 269 2	1.0
	−0.577 350 269 2	1.0
	+0.774 596 697	0.555 555 555 6
3	0.0	0.888 888 888 9
	−0.774 596 697	0.555 555 555 6
4	+0.861 136 311 6	0.347 854 845 1
	−0.861 136 311 6	0.347 854 845 1
	+0.339 981 043 6	0.652 145 154 9
	−0.339 981 043 6	0.652 145 154 9

This table presents the set of Gauss points and weights to integrate numerically polynomial functions up to the seventh order.

In the FE isoparametric formulation, the Jacobian of the transformation is given by

$$J_E(\eta) = \frac{dy}{d\eta} = \sum_{i=1}^{N_{NE}} \frac{dN_i(\eta)}{d\eta} y_i \tag{7.8}$$

where N_{NE} is the number of nodes per element, N_i are the shape functions, and y_i are the node locations. Another useful formula is related to the derivatives of the functions expressed in the global and natural domains:

$$\frac{df(y)}{dy} = \frac{1}{J_E}\frac{df(\eta)}{d\eta} \tag{7.9}$$

Example 7.2.1 *Let us reconsider the data from Example 5.2.1, where the line integrals now have to be computed by means of the Gauss quadrature formula. The shape functions in natural coordinates are given by*

$$N_1 = 1 - \eta,\ N_2 = \eta$$

The nodes of the B2 element are located at $y = 0, L$*, and the Jacobian of the transformation is given by*

$$J = \sum_{i=1}^{2} \frac{dN_i(\eta)}{d\eta} y_i = 1 \times 0 + 1 \times L = L$$

The integrals to be computed are related to the following stiffness matrix component:

$$\begin{aligned}
K_{\mathrm{xx}}^{2122} &= \tilde{C}_{22} \int_{-a}^{+a}\int_{-b}^{+b} F_{2,_x} F_{2,_x} dxdz \int_0^L N_2 N_1 dy \\
&+\tilde{C}_{66} \int_{-a}^{+a}\int_{-b}^{+b} F_{2,_z} F_{2,_z} dxdz \int_0^L N_2 N_1 dy \\
&+\tilde{C}_{44} \int_{-a}^{+a}\int_{-b}^{+b} F_2 F_2\, dxdz \int_0^L N_{2,_y} N_{1,_y} dy
\end{aligned}$$

and in natural coordinates

$$\begin{aligned}
K_{\mathrm{xx}}^{2122} &= \tilde{C}_{22} \int_{-a}^{+a}\int_{-b}^{+b} 1 \cdot 1\, dxdz \int_{-1}^{+1} L\,\eta\,(1-\eta)\, d\eta \\
&+\tilde{C}_{66} \int_{-a}^{+a}\int_{-b}^{+b} 0 \cdot 0\, dxdz \int_{-1}^{+1} L\,\eta\,(1-\eta)\, d\eta \\
&+\tilde{C}_{44} \int_{-a}^{+a}\int_{-b}^{+b} x \cdot x\, dxdz \int_{-1}^{+1} \underbrace{\frac{1}{L}}_{1/J_E} \left(- \underbrace{\frac{1}{L}}_{1/J_E} \right) d\eta
\end{aligned}$$

The polynomials to be integrated are of second order, therefore a second-order quadrature is needed:

$$\int_{-1}^{+1} f(\eta) = f(0) \times 2.0 + f(+0.577\,350\,269\,2) \times 1.0 + f(-0.577\,350\,269\,2) \times 1.0$$

The final result is exactly the same as the analytically computed one:

$$K_{xx}^{2122} = \frac{2}{3}\tilde{C}_{22}\, a\, b\, L - \frac{4}{3}\tilde{C}_{44}\,\frac{b\, a^3}{L}$$

7.2.1.3 Shear locking

Shear locking is a numerical phenomenon that can occur as the thickness of beams (Reddy, 1997) or plates decreases (Cinefra *et al.* 2010). Locking is due to an overestimation of the shear stiffness of the structures, which tends to be infinite as the thickness tends to zero. Many techniques are able to attenuate this effect. Amongst these, the one adopted for the present formulation is the reduced selective integration. This method is based on a reduced Gauss integration of the terms of the stiffness matrix that are related to shear. "Reduced" here means that a lower number of Gaus points are used and this results in a reduction in the shear stiffness of the structure. It is important to point out that refined beam theories obtained by means of the CUF can lead to models which are hardly affected by shear locking, as shown in Example 7.2.2.

Example 7.2.2 *Let us consider the cantilevered beam analyzed in Example 7.1.2. The effect of the selective integration has to be evaluated by using a fourth-order model. Table 7.12 shows the results for different meshes. It is evident that:*

- *The use of a fourth-order model makes the selective integration necessary only if beams are meshed with B2 elements.*
- *In all the other cases, the full integration is able to detect the right solution.*
- *The detrimental effect of shear locking due to low thickness values is confirmed.*
- *The use of higher-order models provide a shear-locking-free model.*

7.2.1.4 Surface integrals

Surface integrals of F_τ are evaluated numerically by partitioning the integration area into a certain number of sub-domains:

$$\int_\Omega F_\tau(x,z)\, F_s(x,z)\, d\Omega \simeq \sum_{m=1}^{M} [F_\tau(x_m, z_m)\, F_s(x_m, z_m)]\, \Omega_m \quad (7.10)$$

where x_m and z_m are the coordinates of the center of Ω_m. M is evaluated through a convergence study. This numerical technique is adopted in order to analyze

Table 7.12 Effect of the selective integration on u_z from different meshes and beam elements via a $N = 4$ model (Carrera *et al.* 2010a).

No. of elements	B2	B2*	B3	B3*	B4	B4*
		$L/h = 100$, $u_y \times 10^2$ m				
1	−0.893	−0.0004	−1.158	−0.905	−1.240	−1.227
2	−1.065	−0.001	−1.255	−1.191	−1.290	−1.278
3	−1.165	−0.003	−1.275	−1.247	−1.302	−1.297
4	−1.210	−0.006	−1.289	−1.271	−1.310	−1.310
5	−1.236	−0.009	−1.298	−1.285	−1.320	−1.312
10	−1.287	−0.035	−1.316	−1.312	−1.325	−1.324
20	−1.311	−0.129	−1.325	−1.324	−1.329	−1.329
40	−1.323	−0.399	−1.330	−1.329	−1.332	−1.332
		$L/h = 10$, $u_y \times 10^5$ m				
1	−0.904	−0.035	−1.168	−0.988	−1.250	−1.241
2	−1.076	−0.128	−1.255	−1.223	−1.296	−1.293
3	−1.176	−0.255	−1.285	−1.274	−1.311	−1.310
4	−1.220	−0.392	−1.299	−1.294	−1.319	−1.318
5	−1.246	−0.524	−1.307	−1.305	−1.323	−1.322
10	−1.296	−0.954	−1.324	−1.323	−1.330	−1.330
20	−1.318	−1.208	−1.330	−1.330	−1.333	−1.333
40	−1.328	−1.298	−1.333	−1.333	−1.333	−1.333

*Full integration adopted.
This tables shows the effect of the selective integration on the solution given by a fourth-order beam model for different meshes.

arbitrary cross-section geometries since analytical solutions of the surface integrals might not be efficient from a numerical point of view. The sub-domains can have arbitrary geometries. A triangular shape is usually preferred since it permits a better partitioning of irregular cross-sections (e.g., annular or airfoil shaped).

The area of each triangular element, Ω_m, can be computed starting from the vertices of the triangle received as input:

$$\Omega_m = \frac{1}{2}\left[\det\begin{pmatrix} x_{1m} & x_{2m} & x_{3m} \\ z_{1m} & z_{2m} & z_{3m} \\ 1 & 1 & 1 \end{pmatrix}\right] \tag{7.11}$$

where the first two rows of the matrix contain the coordinates of the three vertices of the triangle. The center point of each triangle is computed as

$$\begin{aligned} x_m &= \frac{x_{1m} + x_{2m} + x_{3m}}{3} \\ z_m &= \frac{z_{1m} + z_{2m} + z_{3m}}{3} \end{aligned} \tag{7.12}$$

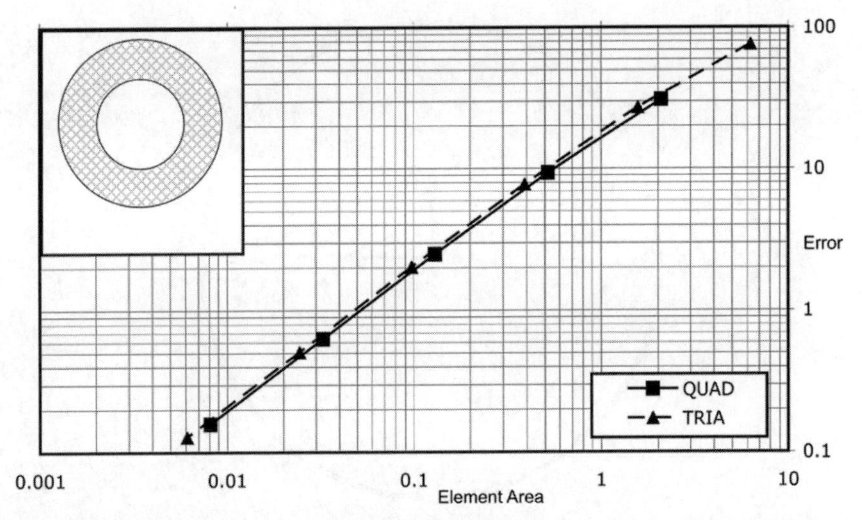

Figure 7.12 Convergence study of $\int_\Omega x^8 d\Omega$ above an annular domain.

Figure 7.12 shows a typical convergence study on a numerical mesh of an annular cross-section. The computed integral is

$$\int_\Omega x^8 d\Omega \tag{7.13}$$

Both quadrilateral and triangular mesh elements are used. The horizontal axis reports the ratio between the mean area of the elements and the total area of the cross-section. The vertical axis shows the error with respect to the exact solution, which, in this case, can be easily obtained.

Example 7.2.3 *This example provides reference numerical integral data for comparison purposes. Two cross-section geometries are considered and different order integrals are computed. The first cross-section is shown in Figure 7.13, the annular cross-section of the outer radius equal to* 2 m, *and the thickness equal to* 0.02 m. *Numerical integrals are computed via a mesh composed of* 10^5 *triangular meshes. The second cross-section is shown in Figure 7.14, where the airfoil-shaped cross-section was built by using the NACA profile 2415 with unit chord and two vertical walls located at 25% and 75% chordwise. Numerical integrals are computed via a mesh composed of* 1.5×10^5 *triangular meshes. Results are given in Tables 7.13 and 7.14.*

7.2.2 Stiffness and mass matrix numerical examples

Some numerical examples of stiffness and mass matrices are provided in this section in order to supply reference data that could be useful for comparison purposes.

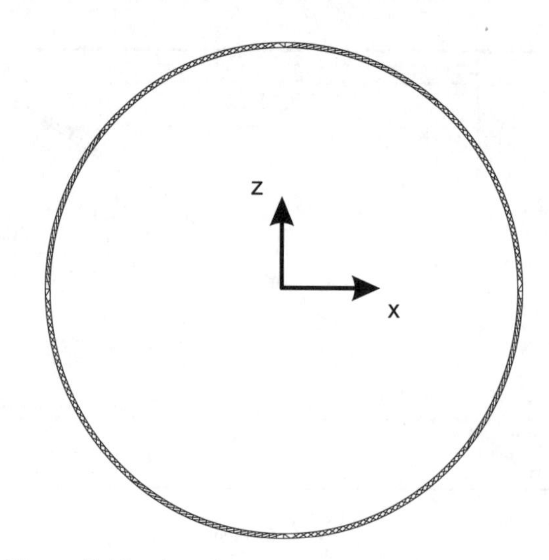

Figure 7.13 Annular cross-section numerical mesh.

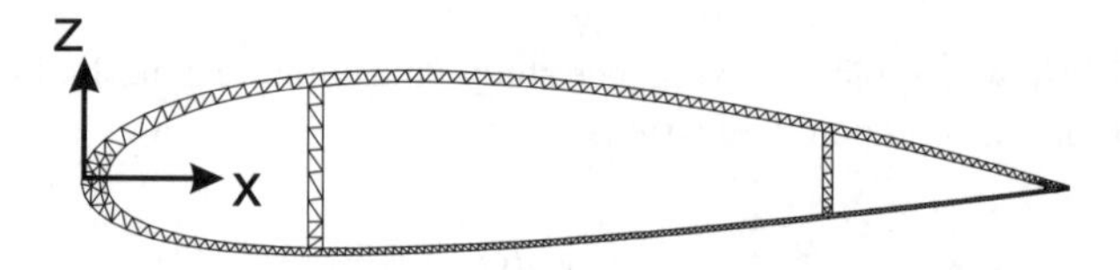

Figure 7.14 Airfoil-shaped cross-section numerical mesh.

The first data set is related to a simply supported beam with an annular cross-section. The outer diameter of the cylinder, d, is equal to 2 m, and $L/d = 10$. The thin-walled structure is 0.02 m thick. A mesh of 10 B4 elements is adopted with an $N = 1$ model. The material is isotropic with E equal to 75 GPa, ν equal to 0.33, and the density, ρ, is 2700 kg/m^3. The surface integrals are computed numerically by means of a 10^5 triangular numerical mesh. Some components of the stiffness

Table 7.13 Numerical integrals of the annular cross-section.

$\int_\Omega$	1	x	z	x^2
1	1.244×10^{-1}	0.0	0.0	6.097×10^{-2}
x	0.0	6.097×10^{-2}	0.0	0.0
z	0.0	0.0	6.097×10^{-2}	0.0
x^2	6.097×10^{-2}	0.0	0.0	4.483×10^{-2}

This table presents a set of cross-section integrals computed numerically; each value is related to the integral of the product between the row and column function.

Table 7.14 Numerical integrals of the airfoil-shaped cross-section.

$\int_\Omega$	x^3	$x^2 z$	$x z^2$	z^3
x^3	5.371×10^{-5}	-1.027×10^{-6}	1.870×10^{-7}	1.856×10^{-9}
$x^2 z$	-1.027×10^{-6}	1.870×10^{-7}	1.856×10^{-9}	5.056×10^{-9}
$x z^2$	1.870×10^{-7}	1.856×10^{-9}	5.056×10^{-9}	4.734×10^{-10}
z^3	1.856×10^{-9}	5.056×10^{-9}	4.734×10^{-10}	7.518×10^{-10}

This table presents a set of cross-section integrals computed numerically, each value is related to the integral of the product between the row and column function.

matrix are reported; in particular the terms related to the first B4 element are considered for $i = j$. The stiffness matrix components considered are then placed between the 10th and 18th rows and columns

	$s = 1$			$s = 2$	$s = 3$
$\tau = 1$	1.894×10^{10}	0.0	0.0	...	...
	0.0	5.039×10^{10}	0.0		
	0.0	0.0	1.894×10^{10}		
$\tau = 2$	...			...	...
$\tau = 3$	...			...	...

$i = j = 2$, 1st element

The first bending frequency for this structural model is equal to 14.182 Hz (Carrera *et al.* 2011).

The mass matrix is now considered for the same structural model, but this time an $N = 2$ theory is used. In this case, components related to second-order terms are considered, $\tau, s = 4, 5, 6$. These terms are placed between the 19th and 36th rows and columns since the first B4 element is again considered with $i = j = 2$

	$s = 4$	$s = 5$			$s = 6$
$\tau = 4$	...	...			...
$\tau = 5$	...	3.112×10^{1}	0.0	0.0	...
		0.0	$3.112 \times 10^{1} \times 10^{10}$	0.0	
		0.0	0.0	3.112×10^{1}	
$\tau = 6$	...	...			...

$i = j = 2$, 1st element

The non-null values have the same value because of the symmetry of the structure; they are in fact given by the following integrals:

$\rho \int_l N_2\, N_2\, dl \int_\Omega x^4 d\Omega$	0.0	0.0
0.0	$\rho \int_l N_2\, N_2\, dl \int_\Omega x^2\, z^2 d\Omega$	0.0
0.0	0.0	$\rho \int_l N_2\, N_2\, dl \int_\Omega z^4 d\Omega$

i=j=2, 1st element

In this case the first bending frequency is equal to 14.185 Hz. A higher frequency value is found than for the $N = 1$ case, that is, a stiffer structure is modeled by the parabolic model. This phenomenon is due to the aforementioned Poisson locking correction effect, which can make linear models less stiffer than second-order ones.

The thin-walled cylinder considered so far will be reconsidered in the following chapters in which the so-called shell-like capabilities of higher-order models are presented. It will be shown how certain shell-like modal shapes will appear from the $N = 3$ model onwards. However, correct natural frequencies related to this kind of mode are detected by the $N = 7$ model (Carrera *et al.* 2010), with τ and s varying from 1 to 36 and with 108 degrees of freedom per node. Some mass matrix values of this model are reported:

	$s = 34$	$s = 35$	$s = 36$
$\tau = 34$	...	...	...
$\tau = 35$	...	3.640 | 0.0 | 0.0 0.0 | 3.640 | 0.0 0.0 | 0.0 | 3.640	...
$\tau = 36$	...	...	...

$i = j = 2$, 1st element

which correspond to

$\rho \int_l N_2\, N_2\, dl \int_\Omega x^4\, z^{10} d\Omega$	0.0	0.0
0.0	$\rho \int_l N_2\, N_2\, dl \int_\Omega x^2\, z^{12} d\Omega$	0.0
0.0	0.0	$\rho \int_l N_2\, N_2\, dl \int_\Omega z^{14} d\Omega$

$i = j = 2$, 1st element

7.2.3 Constraints and reduced models

This section illustrates the technique exploited to impose constraints on the structure and to create reduced higher-order beam models. The methodology is based on a penalty technique (Bathe 1996), which acts on the stiffness matrix. The procedure is quite straightforward:

- The degrees of freedom to be constrained have to be chosen.
- The nodes to be constrained have to be selected.
- All the *diagonal terms* of the selected node stiffness matrix related to the chosen degrees of freedom have to be penalized with a penalty value, Π.

The value to be assigned to Π should be chosen via a convergence study; however, one reliable way of choosing it is to exploit the maximum value of the stiffness matrix, K_{max}:

$$\Pi \geq 10^3 \times K_{max} \tag{7.14}$$

At each diagonal position, K_{ii}, which has to be penalized, the following substitution has to be made:

$$K_{ii} \rightarrow \Pi$$

While constraints act on a set of nodes, if reduced models have to be used, all the nodes will be affected by the penalty (Carrera and Petrolo, 2011). This means that to obtain a displacement model, such as

$$\begin{aligned} u_x &= x\, u_{x_2} + x^2\, u_{x_4} + z^2\, u_{x_6} \\ u_y &= u_{y_1} + x\, u_{y_2} + xz\, u_{y_5} \\ u_z &= x\, u_{z_2} + z\, u_{z_3} + xz\, u_{z_5} \end{aligned}$$

the following procedure is needed:

1. A second-order beam model has to be implemented.
2. The diagonal terms corresponding to the generalized variables u_{x_1}, u_{x_3}, u_{x_5}, u_{y_3}, u_{y_4}, u_{y_6}, u_{z_1}, u_{z_4}, and u_{z_6} have to be penalized in all the nodes of the FE model.

Example 7.2.4 *Let us consider a node of an $N = 1$ model and three different constraint options: clamped, hinged, and hinged with the horizontal translation allowed. The proper positions of the penalty value, Π, within the nodal stiffness matrix have to be calculated for all the three constraint cases considered. A full linear model has nine generalized displacement variables per node, that* *is, the nodal stiffness matrix is a 9×9 array. Penalties have to be inserted on diagonal terms only, thus $i = j$ and $\tau = s$. In the case of a clamped node, all the displacement variables have to be penalized*

$$
\begin{array}{c||c|c|c}
 & s=1 & s=2 & s=3 \\
\hline\hline
\tau=1 & \begin{array}{c|c|c} \Pi & & \\ \hline & \Pi & \\ \hline & & \Pi \end{array} & \cdots & \cdots \\
\hline
\tau=2 & \cdots & \begin{array}{c|c|c} \Pi & & \\ \hline & \Pi & \\ \hline & & \Pi \end{array} & \cdots \\
\hline
\tau=3 & \cdots & \cdots & \begin{array}{c|c|c} \Pi & & \\ \hline & \Pi & \\ \hline & & \Pi \end{array}
\end{array}_{\,i=j}
$$

If the node has to be hinged, the rotations with respect to the x- and z-axis has to be unconstrained

$$
\begin{array}{c||c|c|c}
 & s=1 & s=2 & s=3 \\
\hline\hline
\tau=1 & \begin{array}{c|c|c} \Pi & & \\ \hline & \Pi & \\ \hline & & \Pi \end{array} & \cdots & \cdots \\
\hline
\tau=2 & \cdots & \begin{array}{c|c|c} \Pi & & \\ \hline & & \\ \hline & & \Pi \end{array} & \cdots \\
\hline
\tau=3 & \cdots & \cdots & \begin{array}{c|c|c} \Pi & & \\ \hline & & \\ \hline & & \Pi \end{array}
\end{array}_{\,i=j}
$$

If a rolling supported node has to be implemented, the translation along the axial direction has to be left free as well

$$
\begin{array}{c||c|c|c}
 & s=1 & s=2 & s=3 \\
\hline\hline
\tau=1 & \begin{array}{c|c|c} \Pi & & \\ \hline & & \\ \hline & & \Pi \end{array} & \dots & \dots \\
\hline
\tau=2 & \dots & \begin{array}{c|c|c} \Pi & & \\ \hline & & \\ \hline & & \Pi \end{array} & \dots \\
\hline
\tau=3 & \dots & \dots & \begin{array}{c|c|c} \Pi & & \\ \hline & & \\ \hline & & \Pi \end{array}
\end{array}
\quad {}_{i=j}
$$

Example 7.2.5 *Let us consider the following reduced beam model:*

$$
\begin{aligned}
u_x &= x\, u_{x_2} + z\, u_{x_3} \\
u_y &= u_{y_1} + x\, u_{y_2} + z\, u_{y_3} \\
u_z &= x\, u_{z_2} + z\, u_{z_3}
\end{aligned}
$$

It has to be implemented via the penalty technique. A first-order model has to be chosen, therefore the nodal stiffness matrix is a 9×9 array. All the diagonal terms related to the variables u_{x_1} and u_{z_1} have to be penalized for all the nodes

$$
\begin{array}{c||c|c|c}
 & s=1 & s=2 & s=3 \\
\hline\hline
\tau=1 & \begin{array}{c|c|c} \Pi & & \\ \hline & & \\ \hline & & \end{array} & \dots & \dots \\
\hline
\tau=2 & \dots & \begin{array}{c|c|c} & & \\ \hline & & \\ \hline & & \end{array} & \dots \\
\hline
\tau=3 & \dots & \dots & \begin{array}{c|c|c} \Pi & & \\ \hline & & \\ \hline & & \end{array}
\end{array}
\quad {}_{i=j,\ \forall \text{ nodes}}
$$

This operation can easily be made automatically by acting on the FE code where an "if" structure similar to the following has to be inserted:

```
IF (i == j) .AND. ( tau == s) .AND.
& ( tau == 1 .OR. tau == 3) THEN
    K(1,1) = PI
ENDIF
```

7.2.4 Load vector

The computation of the equivalent load nodal vector is another issue where the use of CUF models makes specific tasks relevant. Higher-order terms play a fundamental role in determining this vector. As in a general FE model, some steps have to be taken in order to construct the force vector:

1. The type of load has to be chosen (e.g., concentrated, surface, inertial load, etc.).
2. The portion of the structure to be loaded has to be indicated in terms of coordinate boundaries.
3. The elements, and then the nodes involved in the loading region, have to be identified.
4. The loading vector is then built by means of the PVD.

The aim of this section is to underline the role of higher-order terms in the loading vector. Example 7.2.6 presents a comprehensive set of loading cases that are able to highlight the importance of refined theories.

Example 7.2.6 *Let us consider a B2 element having a square cross-section and loaded with two opposite forces of equal magnitude as shown in Figures 7.15 and 7.16. The equivalent nodal vector has to be computed for TBT and* $N = 1$ *models. The load vector is computed by exploiting PVD:*

$$\delta L_{ext} = \boldsymbol{P}\,\delta\boldsymbol{u} = \boldsymbol{P}\,F_\tau\,N_i\,\boldsymbol{q}_{\tau i} \tag{7.15}$$

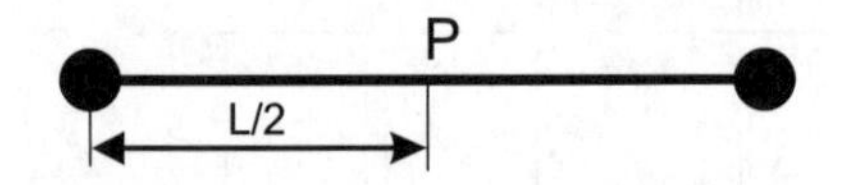

Figure 7.15 B2 element loaded at $L/2$.

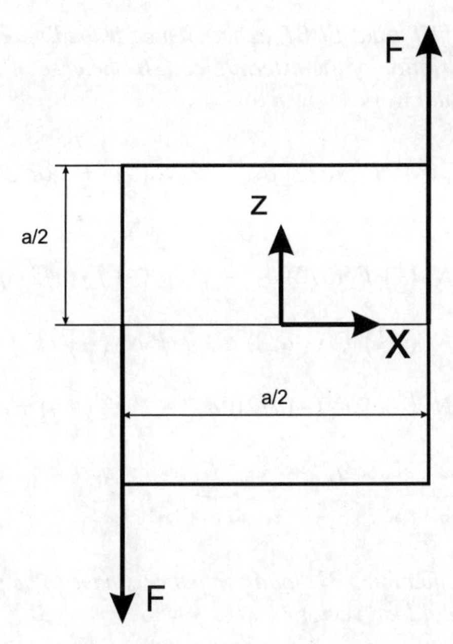

Figure 7.16 B2 element square cross-section torsion load.

Shape functions, N_i, at the loading point are equal to

$$N_1\left(\frac{1}{2}\right) = N_2\left(\frac{1}{2}\right) = \frac{1}{2}$$

P acts along the z-direction, that is, only the q_z components of the loading vector are involved. In the case of TBT, the virtual variation of the virtual work is given by

$$\begin{aligned}
\delta L_{\text{ext}} &= P\,N_1\left(\frac{1}{2}\right)\underbrace{F_1}_{F_1=1}(a/2)\,q_{z_{11}} + P\,N_2\left(\frac{1}{2}\right)F_1\,(a/2)\,q_{z_{12}} \\
&\quad - P\,N_1\left(\frac{1}{2}\right)F_1(-a/2)\,q_{z_{11}} - P\,N_2\left(\frac{1}{2}\right)F_1\,(-a/2)\,q_{z_{12}} \\
&= \frac{1}{2}P\,q_{z_{11}} + \frac{1}{2}P\,q_{z_{12}} - \frac{1}{2}P\,q_{z_{11}} - \frac{1}{2}P\,q_{z_{12}} \\
&= 0
\end{aligned}$$

It is clear that by TBT (and EBBT as well) such a loading condition does not involve a virtual variation of the external load. In the case of $N = 1$, the virtual variation of the virtual work is given by

$$\begin{aligned}
\delta L_{ext} = & P\, N_1\left(\tfrac{1}{2}\right) \underbrace{F_1}_{F_1=1} (a/2)\, q_{z_{11}} + P\, N_1\left(\tfrac{1}{2}\right) \underbrace{F_2}_{F_2=x} (a/2)\, q_{z_{21}} \\
& + P\, N_2\left(\tfrac{1}{2}\right) F_1\,(a/2)\, q_{z_{12}} + P\, N_2\left(\tfrac{1}{2}\right) F_2\,(a/2)\, q_{z_{22}} \\
& - P\, N_1\left(-\tfrac{1}{2}\right) F_1\,(-a/2)\, q_{z_{11}} - P\, N_1\left(-\tfrac{1}{2}\right) F_2\,(-a/2)\, q_{z_{21}} \\
& - P\, N_2\left(-\tfrac{1}{2}\right) F_1\,(-(a/2))\, q_{z_{12}} - P\, N_2\left(-\tfrac{1}{2}\right) F_2\,(-a/2)\, q_{z_{22}} \\
= & \underbrace{\tfrac{1}{2}\, P\, a}_{\text{First node load}} \times q_{z_{21}} + \underbrace{\tfrac{1}{2}\, P\, a}_{\text{Second node load}} \times q_{z_{22}}
\end{aligned}$$

and, according to Equation 5.31, load components have to be placed in the sixth position of each nodal load vector.

7.3 Postprocessing

The result from postprocessing represents the final step of a FE analysis. In a displacement-based model, the postprocessing input is the vector of the generalized nodal displacement variables, for the case of a static analysis, or the eigenvectors, for the case of a modal analysis. These vectors need to be expanded over all the points where postprocessing data are needed. The expansion process involves shape functions, $N_i(y)$, along the beam axis and expansion functions, F_τ, above the cross-section

$$\left\{\begin{array}{l}
q_x(x_p, y_p, z_p) = N_i(y_p)\, F_\tau(x_p, z_p)\, q_{x_{\tau i}} \\
q_y(x_p, y_p, z_p) = N_i(y_p)\, F_\tau(x_p, z_p)\, q_{y_{\tau i}} \\
q_z(x_p, y_p, z_p) = N_i(y_p)\, F_\tau(x_p, z_p)\, q_{x_{\tau i}}
\end{array}\right. \tag{7.16}$$

where $[x_p, y_p, z_p]$ is a generic point of the structure in which the displacement components have to be computed. The terms $q_{x_{\tau i}}$ have to be extracted from the nodal unknown vector; the extraction process requires that the FE to which the postprocessing point belongs is identified.

Example 7.3.1 *Let us consider the B2 element in Figure 7.15. The displacement component q_x has to be computed in P in a generic cross-section point of* *coordinates $[x_p, z_p]$. An $N = 2$ model is assumed:*

$$q_x(x_p, y_p, z_p) = N_1(y_p)\, F_\tau(x_p, z_p)\, q_{x_{\tau 1}} + N_2(y_p)\, F_\tau(x_p, z_p)\, q_{x_{\tau 2}}$$

where $q_{x_{\tau 1}}$ and $q_{x_{\tau 2}}$ are the x components of the nodal variables of the first and second nodes of the element respectively. Shape functions in P are equal to 1/2:

$$\begin{aligned} q_x(x_p, y_p, z_p) = \frac{1}{2} \big[& F_1(x_p, z_p)\, q_{x_{11}} + F_2(x_p, z_p)\, q_{x_{21}} \\ & + F_3(x_p, z_p)\, q_{x_{31}} + F_4(x_p, z_p)\, q_{x_{41}} \\ & + F_5(x_p, z_p)\, q_{x_{51}} + F_6(x_p, z_p)\, q_{x_{61}} \big] \\ + \frac{1}{2} \big[& F_1(x_p, z_p)\, q_{x_{12}} + F_2(x_p, z_p)\, q_{x_{22}} \\ & + F_3(x_p, z_p)\, q_{x_{32}} + F_4(x_p, z_p)\, q_{x_{42}} \\ & + F_5(x_p, z_p)\, q_{x_{52}} + F_6(x_p, z_p)\, q_{x_{62}} \big] \end{aligned}$$

Substituting the F_τ expressions, the displacement component becomes

$$\begin{aligned} q_x(x_p, y_p, z_p) = \frac{1}{2} [& q_{x_{11}} + x_p\, q_{x_{21}} + z_p\, q_{x_{31}} \\ & + x_p^2\, q_{x_{41}} + x_p\, z_p\, q_{x_{51}} + z_p^2\, q_{x_{61}}] \\ + \frac{1}{2} [& q_{x_{12}} + x_p\, q_{x_{22}} + z_p\, q_{x_{32}} \\ & + x_p^2\, q_{x_{42}} + x_p\, z_p\, q_{x_{52}} + z_p^2\, q_{x_{62}}] \end{aligned}$$

Modal shapes from free-vibration analyses have to be computed by considering each eigenvector as equivalent to a nodal displacement vector and then expanding it as in Example 9.

7.3.1 Stresses and strains

Stress and strain components can be computed straightforwardly from the nodal displacement vector. Strains, in particular, require that the partial derivatives are calculated by means of shape and expansion functions:

$$\begin{cases} \frac{\partial \mathbf{q}}{\partial x} = \frac{\partial (F_\tau \, N_i \, \mathbf{q}_{\tau i})}{\partial x} = \frac{\partial (F_\tau)}{\partial x} N_i \, \mathbf{q}_{\tau i} = F_{\tau,x} \, N_i \, \mathbf{q}_{\tau i} \\ \frac{\partial \mathbf{q}}{\partial y} = \frac{\partial (F_\tau \, N_i \, \mathbf{q}_{\tau i})}{\partial y} = \frac{\partial (N_i)}{\partial x} F_\tau \, \mathbf{q}_{\tau i} = F_\tau \, N_{i,y} \, \mathbf{q}_{\tau i} \\ \frac{\partial \mathbf{q}}{\partial z} = \frac{\partial (F_\tau \, N_i \, \mathbf{q}_{\tau i})}{\partial z} = \frac{\partial (F_\tau)}{\partial z} N_i \, \mathbf{q}_{\tau i} = F_{\tau,z} \, N_i \, \mathbf{q}_{\tau i} \end{cases} \tag{7.17}$$

As the strain field is computed, stress components are obtained through constitutive laws:

$$\{\boldsymbol{\sigma}\} = [\mathbf{C}] \, \{\boldsymbol{\epsilon}\}$$

Example 7.3.2 *Let us consider Example 7.3.1 to compute ϵ_{xx}:*

$$\epsilon_{xx}(x_p, y_p, z_p) = \frac{\partial q_x(x_p, y_p, z_p)}{\partial x} = N_1(y_p) \, F_{\tau,x}(x_p, z_p) \, q_{x_{\tau 1}} + N_2(y_p) \, F_{\tau,x}(x_p, z_p) \, q_{x_{\tau 2}}$$

thus

$$\begin{aligned} \epsilon_{xx}(x_p, y_p, z_p) = \frac{1}{2} & \left[F_{1,x}(x_p, z_p) \, q_{x_{11}} + F_{2,x}(x_p, z_p) \, q_{x_{21}} \right. \\ & + F_{3,x}(x_p, z_p) \, q_{x_{31}} + F_{4,x}(x_p, z_p) \, q_{x_{41}} \\ & \left. + \; F_{5,x}(x_p, z_p) \, q_{x_{51}} + F_{6,x}(x_p, z_p) \, q_{x_{61}} \right] \\ & + \frac{1}{2} \left[F_{1,x}(x_p, z_p) \, q_{x_{12}} + F_{2,x}(x_p, z_p) \, q_{x_{22}} \right. \\ & + F_{3,x}(x_p, z_p) \, q_{x_{32}} + F_{4,x}(x_p, z_p) \, q_{x_{42}} \\ & \left. + \; F_{5,x}(x_p, z_p) \, q_{x_{52}} + F_{6,x}(x_p, z_p) \, q_{x_{62}} \right] \end{aligned}$$

Substituting the $F_{\tau,x}$ expressions, the displacement component becomes

$$\begin{aligned} q_x(x_p, y_p, z_p) = & \frac{1}{2} \left[q_{x_{21}} + 2 \, x_p \, q_{x_{41}} + z_p \, q_{x_{51}} \right] \\ & + \frac{1}{2} \left[q_{x_{22}} + 2 \, x_p \, q_{x_{42}} + z_p \, q_{x_{52}} \right] \end{aligned}$$

References

Bathe K 1996 *Finite element procedure*. Prentice Hall.

Carrera E and Nali P 2009 Mixed piezoelectric plate elements with direct evaluation of transverse electric displacement. *International Journal for Numerical Methods in Engineering*, **80**(4), 403–424.

Carrera E and Petrolo M 2011 On the effectiveness of higher-order terms in refined beam theories. *Journal of Applied Mechanics*, **78**(2), DOI: 10.1115/1.4002207.

Carrera E, Giunta G, Nali P, and Petrolo M 2010a Refined beam elements with arbitrary cross-section geometries. *Computers & Structures*, **88**(5–6), 283–293. DOI: 10.1016/j.compstruc.2009.11.002.

Carrera E, Petrolo M, and Nali P 2011 Unified formulation applied to free vibrations finite element analysis of beams with arbitrary section. *Shock and Vibrations*, **18**(3), 485–502. DOI: 10.3233/SAV-2010–0528.

Carrera E, Giunta G, and Petrolo M 2010c A modern and compact way to formulate classical and advanced beam theories. *In: Developments in Computational Structures Technology*, Ch. 4. DOI: 10.4203/csets.25.4.

Cinefra M, Carrera E, and Nali P 2010 MITC technique extended to variable kinematic multilayered plate elements. *Composite Structures*, **92**, 1888–1895.

Oñate E 2009 *Structural Analysis with the Finite Element Method: Linear Statics, Volume 1*. Springer.

Reddy, JN 1997 On locking-free shear deformable beam finite elements. *Computer Methods in Applied Mechanics and Engineering*, **149**, 113–132.

References

[illegible] (2009) [illegible] piezo[illegible] plate elements with [illegible] electric [illegible] mechanical [illegible] *Engineering* [illegible] 59 [illegible] 1120.

Carrera E. and Petrolo M. 2012 On the effectiveness of [illegible] of higher-order terms in refined [illegible] theories. *J. Applied Mechanics* 78, DOI: 10.1115/1.[illegible]

Carrera E., [illegible] and Petrolo M. 2010. Refined [illegible] with [illegible] *Composite Structures* [illegible] 93 [illegible] DOI: [illegible]

Carrera E., [illegible] (2011) [illegible] *Mechanics of Advanced Materials and Structures* [illegible] DOI: 10.1[illegible]

Carrera E., [illegible] and Petrolo M. 2011. [illegible] *Thin-Walled Structures* [illegible]

[illegible] M., Carrera E. and [illegible] 2010 [illegible] *Computers and Structures* [illegible]

[illegible] 2009 [illegible]

[illegible] (19[illegible]) On [illegible] elements (Carrera [illegible]). *Mechanics* [illegible]

[illegible] as follows:

8

Shell capabilities of refined beam theories

One of the most important features of a refined beam theory is the possibility of dealing with shell-like structures such as thin-walled bodies. The analysis of this kind of structures requires a number of non-classical effects in the beam model: warping, cross-section in-plane distortion, shear effects, etc. These effects are particularly important when concentrated loads or free vibrations are considered. This chapter is devoted to the analysis of a number of thin shell-like sections to show the enhanced capabilities of the present 1D models and to highlight the advantageous computational costs of the present formulation compared to 2D and 3D elements.

8.1 C-shaped cross-section and bending–torsional loading

Beams with a C-shaped cross-section are subjected to a uniform loading that acts as shown in Figure 8.1. Closed form solutions are considered. Table 8.1 presents the out-of-plane displacement components for $L/a = 100$ and 10. The results are non-dimensionalized as follows:

$$\left(\overline{u}_x, \overline{u}_y, \overline{u}_z\right) = \frac{aE}{L^2 p_{xx}^{+1}} \left(u_x, u_y, u_z\right) \tag{8.1}$$

Beam Structures: Classical and Advanced Theories, First Edition. Erasmo Carrera, Gaetano Giunta and Marco Petrolo.

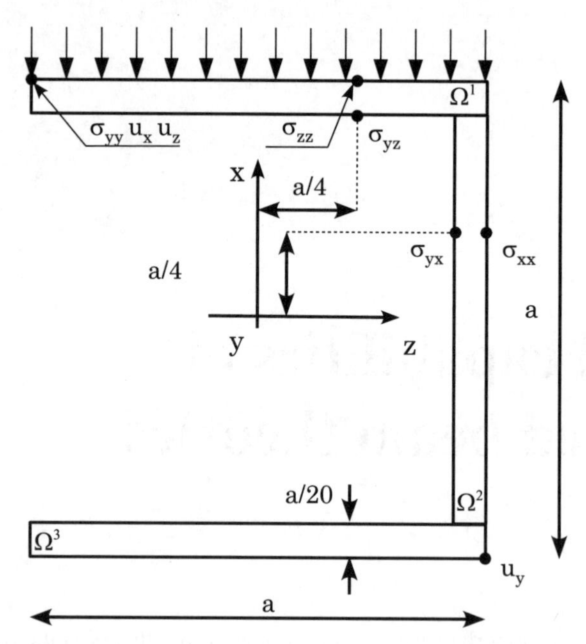

Figure 8.1 C-shaped cross-section geometry, loading and resulting verification points.

Table 8.1 Out-of-plane displacements for a C-shaped beam, $L/a = 100$ and 10 (Giunta *et al.* 2011).

	$L/a = 100$ $10^{-1} \times \overline{u}_y$	$L/a = 10$ $10^{-1} \times \overline{u}_y$
FEM 3D	−9.945	−3.528
$N = 15$	−9.829	−3.502
$N = 8$	−9.277	−3.293
$N = 5$	−8.304	−1.830
$N = 3$	−8.138	−0.878
$N = 1$	−8.131	−0.820
TBT	−8.131	−0.813
EBBT	−8.131	−0.813

This table presents the out-of-plane displacement values of a C-shaped beam obtained through different beam theories and compared to a solid-based model.

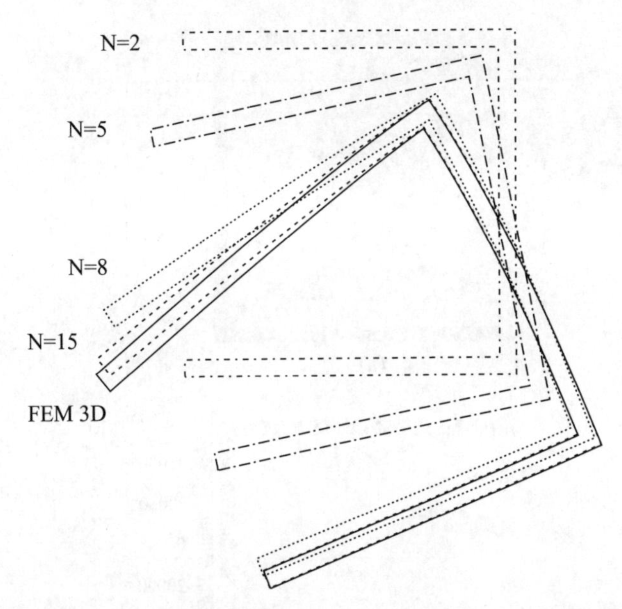

Figure 8.2 In-plane warping of C-shaped beam cross-section at mid-span, $L/a = 10$ (Giunta *et al.* 2011).

The deformed mid-span cross-section for $L/a = 10$ is shown in Figure 8.2. Maps of the stress components are presented in Figures 8.3 and 8.4. For the sake of brevity, only deep beams are considered. The following considerations hold:

1. An accurate prediction of the out-of-plane displacement u_y calls at least for an eighth-order model.

2. The 15th-order model accurately describes in-plane warping. The solution only differs from the FEM one near the free end of the upper branch.

3. A 15th-order model has been considered. Apart from a stress concentration corresponding to the internal corner points, the results are very similar. High gradients of σ_{yx} are modeled correctly.

8.2 Thin-walled hollow cylinder

A thin-walled cylinder is considered and its cross-section geometry is shown in Figure 8.5 where the diameter, d, is equal to 2 m, and the thickness, t, is equal to 0.02 m. The length of the cylinder, L, is equal to 20 m. The structure is modeled as a clamped–clamped beam made of isotropic material ($E = 75$ GPa and $\nu = 0.33$).

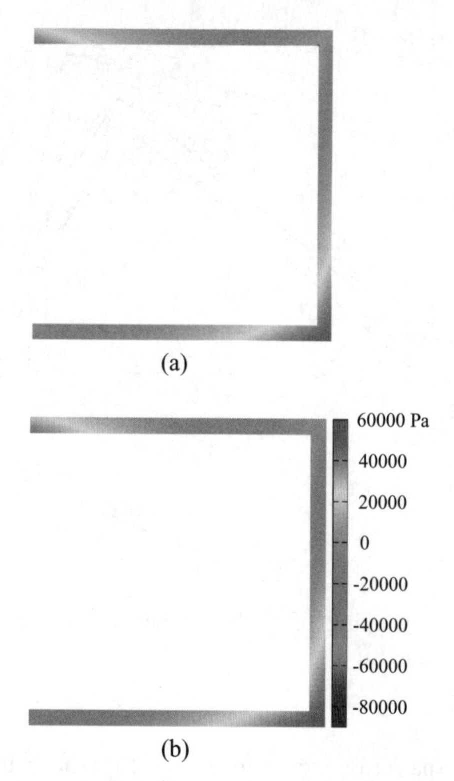

Figure 8.3 Dimensionalized stress σ_{yy} (Pa) above the cross-section for C-shaped beam at mid-span via (a) FEM solution and (b) 15th-order model, $L/a = 10$ (Giunta *et al.* 2011).

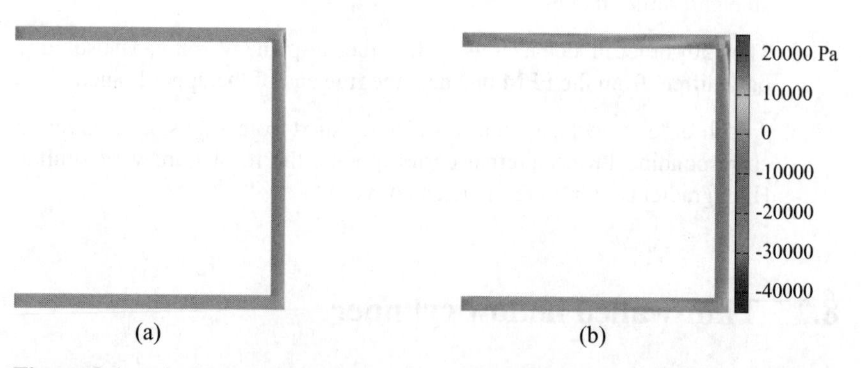

Figure 8.4 Dimensionalized stress σ_{yx} (Pa) above the cross section for C-shaped beam at $y = 0$ via (a) FEM solution and (b) 15th-order model, $L/a = 10$ (Giunta *et al.* 2011).

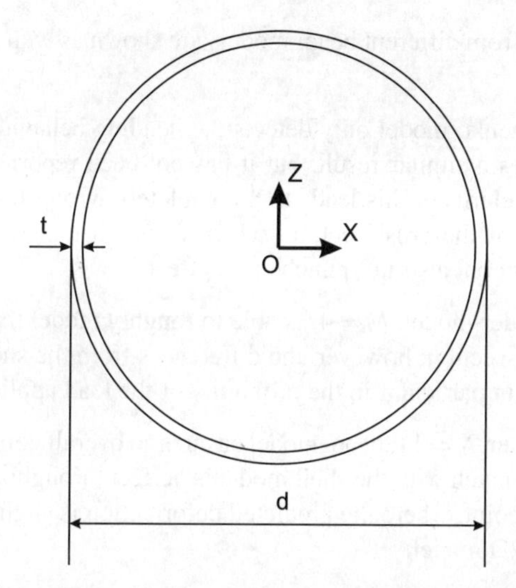

Figure 8.5 Cross-section of the hollow cylinder.

8.2.1 Static analysis: detection of local effects due to a point load

The static analysis is conducted by applying a downward concentrated load, F_z, at $[0, L/2, d/2]$. F_z is parallel to the z-axis and its magnitude is equal to -5 MN. A 10-element B4 mesh is adopted since it leads to convergent results in terms of vertical displacement of the loaded point. An MSC Nastran shell model is used for comparison purposes. Figure 8.6 shows the deformed cross-section at

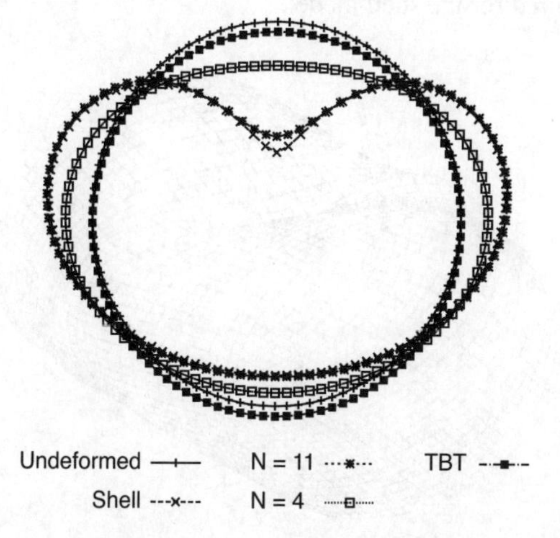

Figure 8.6 Deformed cross-section of the hollow cylinder.

$y = L/2$. Results from different beam models are shown as well as those from the shell model:

- The Timoshenko model only detects the bending behavior of the structure (EBBT gives a similar result, but it has not been reported on the plot for the sake of clarity). This leads to a completely wrong result, especially at the bottom of the cross-section, which does not experience a downward displacement but a sort of punched-up effect.
- A fourth-order model, $N = 4$, is able to roughly model the global behavior of the cross-section; however, the differences from the shell model are still significant, in particular in the proximity of the load application point.
- An 11th-order, $N = 11$, beam model ensures an overall good accuracy for the result. The match with the shell model is perfect throughout but not close to the loaded point, where the predicted deformation is slightly different from that of the 2D model.

Figures 8.7 and 8.8 show the deformed $N = 11$ configuration of the whole cylinder in a 3D and a lateral view, respectively. It can be confirmed that the refined beam model is able to detect accurate 3D deformed states. Table 8.2 presents the u_z values of the loading point for an increasing order of beam models; an indication of the total number of the degrees of freedom of each model is also given. Thus:

- Theories up to second order, $N \leq 2$, only detect the bending behavior.
- Significant improvements in the solution can be observed by increasing the order of the theory, that is, the adoption of higher-order models is effective in detecting the shell-like solution.
- The computational costs of the beam models remain significantly lower than those required for the shell model.

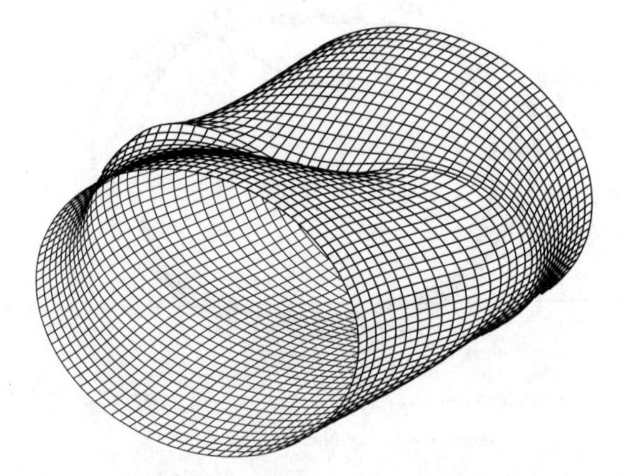

Figure 8.7 The 3D deformation field of the thin-walled cylinder via $N = 11$ beam model.

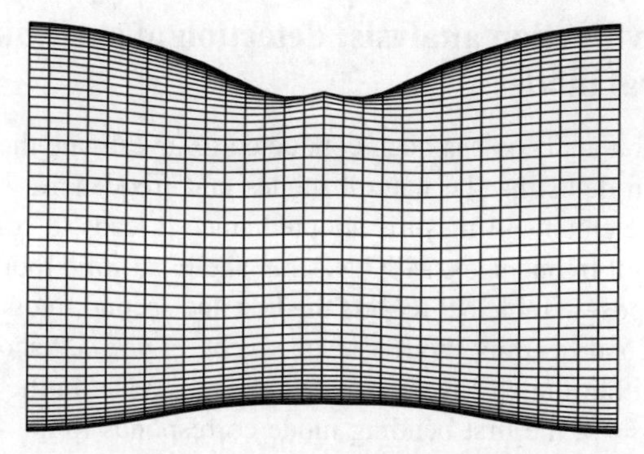

Figure 8.8 The 3D deformation field of the thin-walled cylinder via $N = 11$ beam model, lateral view.

Table 8.2 u_z at the loading point of the thin–walled cylinder (Carrera *et al.* 2010a).

Theory	DOFs	u_z (m)
Detection of the bending behavior only		
EBBT	155	−0.046
TBT	155	−0.053
$N = 1$	279	−0.053
$N = 2$	558	−0.052
Distortion of the cross-section		
$N = 3$	930	−0.114
$N = 4$	1395	−0.229
$N = 5$	1953	−0.335
$N = 6$	2604	−0.386
$N = 7$	3348	−0.486
$N = 8$	4185	−0.535
$N = 9$	5115	−0.564
$N = 10$	6138	−0.584
$N = 11$	7254	−0.597
Shell	49 500	−0.670

This table presents the vertical displacement of the loading point of the thin-walled cylinder; results from different higher-order beam theory and shell models are reported.

8.2.2 Free-vibration analysis: detection of shell-like natural modes

The free-vibration analysis has been conducted to investigate the role of higher-order theories in detecting the natural modes and frequencies of a thin-walled structure. A 10-element B4 mesh is adopted since it leads to convergent results in terms of natural frequencies. MSC Nastran shell and solid models are used for comparison purposes. Table 8.3 reports the first and second frequencies related to the first and second bending modes, respectively. Figure 8.9 shows the position of the modes within the eigenvector matrix. For instance, in the case of a third-order model, $N = 3$, the first bending mode corresponds to the first two natural modes, whereas the second bending mode corresponds to the fifth and sixth natural modes. Each bending mode appears twice in the eigenvector matrix, because of the symmetry of the structure. Thus:

- At least a third-order beam model is needed to obtain a good accuracy for the first two flexural frequencies; moreover, the higher the mode number, the more inaccurate the classical models.
- The use of refined beam theories permits us to obtain a number of modal shapes that are different from the bending ones. In particular, a sixth-order, $N = 6$, beam model is able to correctly detect the first two bending frequencies and also to detect all the natural modes which lie in between the flexural ones.
- The computational costs of the refined beam models remain lower than those of shell and solid elements.

Table 8.3 First and second bending natural frequencies of the thin-walled cylinder (Carrera *et al.* 2010a).

Theory	DOFs	f_1 (Hz)	f_2 (Hz)
EBBT	155	32.598	88.072
TBT	155	30.304	76.447
$N = 1$	279	30.304	76.447
$N = 2$	558	30.730	77.338
$N = 3$	930	28.754	69.448
$N = 4$	1395	28.747	69.402
$N = 5$	1953	28.745	69.397
$N = 6$	2604	28.745	69.397
Shell	49 500	28.489	68.940
Solid	174 000	28.369	68.687

This table presents the frequency values related to the first two bending modes of the thin-walled cylinder.

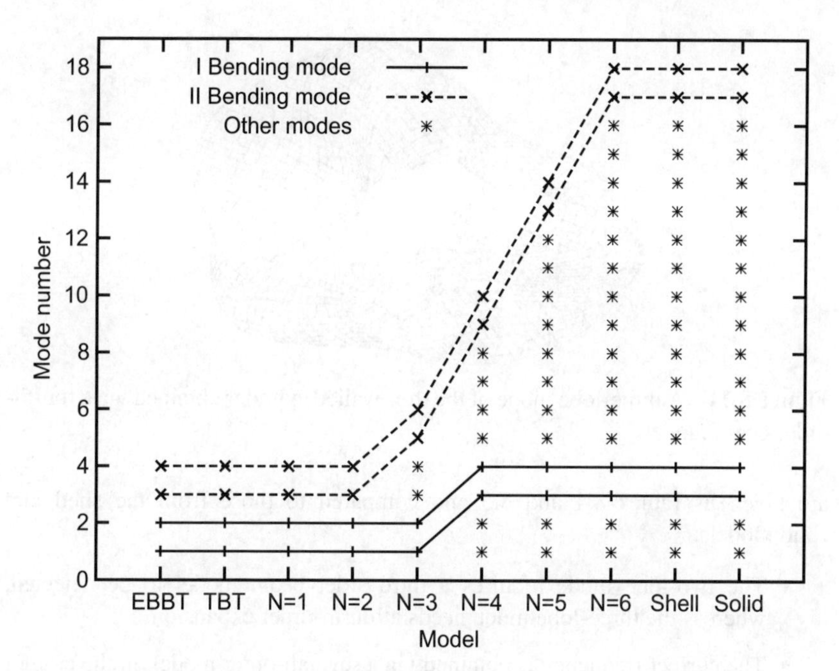

Figure 8.9 Mode-type distribution of the thin-walled cylinder for different structural models.

One type of modal shape that is not detected by lower-order beam models involves the presence of lobes along the circumferential direction of the cylinder. Figures 8.10 and 8.11 show a two- and three-lobe mode, respectively; the modal shapes above the cross-section are presented in Figures 8.12 and 8.13. The frequency values of the first two- and three-lobe frequencies

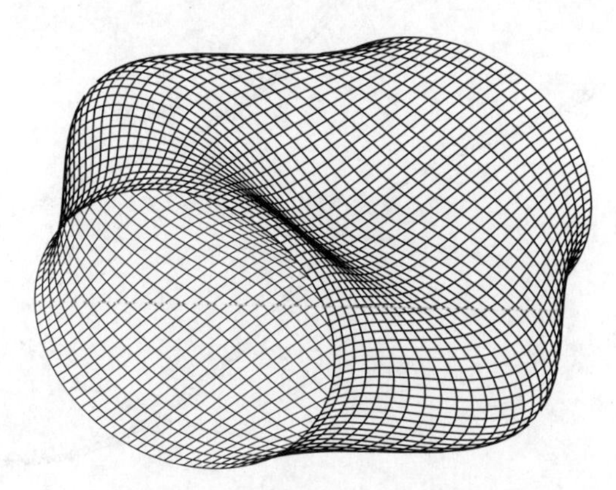

Figure 8.10 A two-lobe mode of the thin-walled cylinder obtained via a third-order beam model.

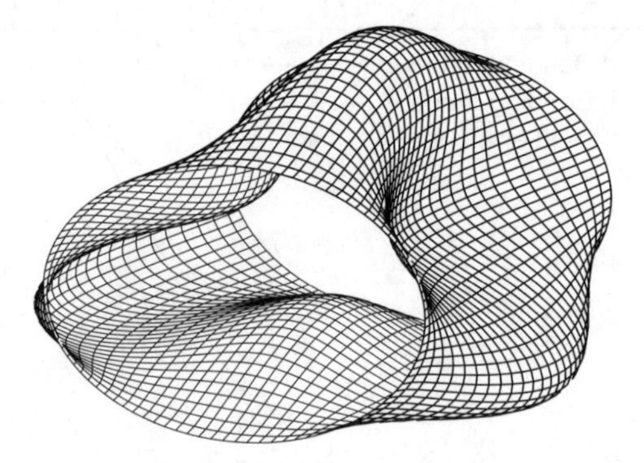

Figure 8.11 A three-lobe mode of the thin-walled cylinder obtained via a fourth-order beam model.

are given in Tables 8.4 and 8.5 and compared to those from the shell and solid models:

- The two-lobe mode requires a third-order beam model to be detected, whereas the three-lobe mode needs a fourth-order expansion.
- The correct frequency is obtained via a seventh-order model, in the case of the two-lobe model, and an eighth-order model must be used to compute the exact frequency.

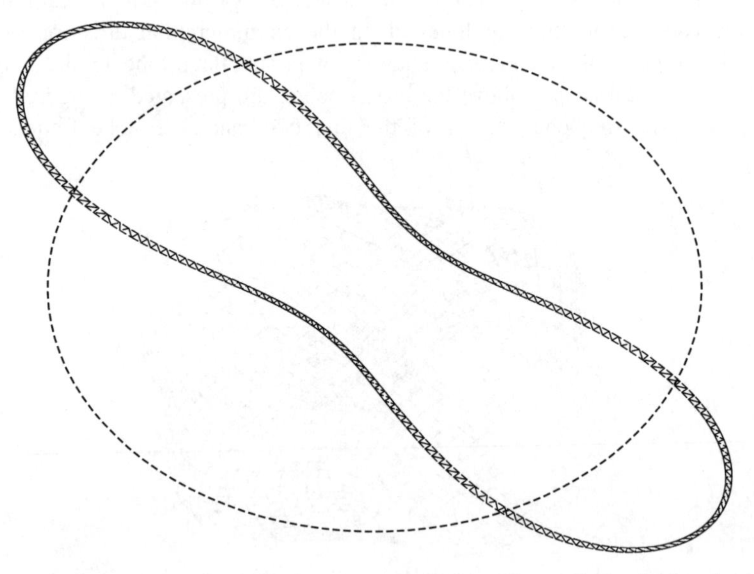

Figure 8.12 The 2D view of a two-lobe mode of the thin-walled cylinder obtained via a third-order beam model.

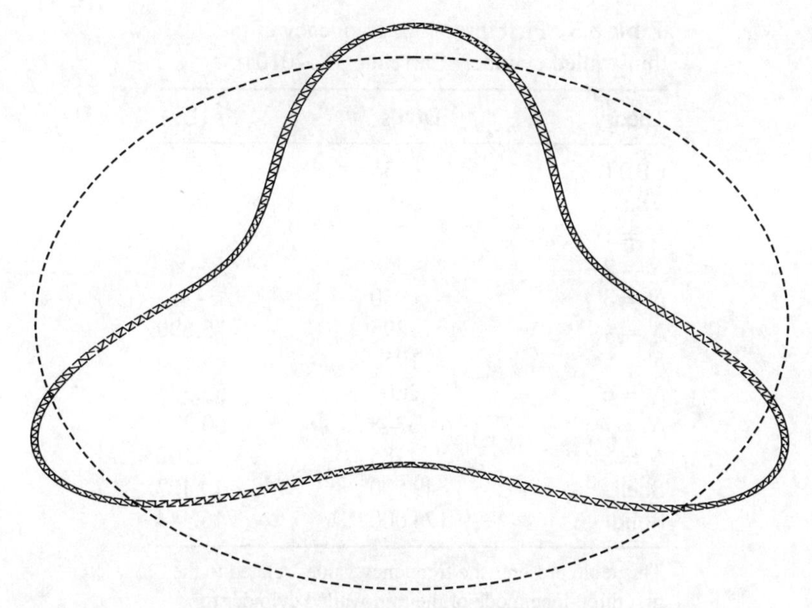

Figure 8.13 The 2D view of a three-lobe mode of the thin-walled cylinder obtained via a fourth-order beam model.

Table 8.4 First two-lobe frequency of the thin-walled cylinder (Carrera *et al.* 2010a).

Theory	DOFs	f (Hz)
EBBT	155	—
TBT	155	—
$N=1$	279	—
$N=2$	558	—
$N=3$	930	38.755
$N=4$	1395	25.156
$N=5$	1953	20.501
$N=6$	2604	20.450
$N=7$	3348	17.363
Shell	49 500	17.406
Solid	174 000	18.932

This table presents the frequency values related to the first two-lobe mode of the thin-walled cylinder for different beam models.

Table 8.5 First three-lobe frequency of the thin-walled cylinder (Carrera *et al.* 2010a).

Theory	DOFs	f (Hz)
EBBT	155	—
TBT	155	—
$N = 1$	279	—
$N = 2$	558	—
$N = 3$	930	—
$N = 4$	1395	75.690
$N = 5$	1953	65.186
$N = 6$	2604	52.386
$N = 7$	3348	50.372
$N = 8$	4185	40.102
Shell	49 500	40.427
Solid	174 000	46.444

This table presents the frequency values related to the first three-lobe mode of the thin-walled cylinder for different beam models.

8.3 Static and free-vibration analyses of an airfoil-shaped beam

An airfoil-shaped cross-section is considered, as shown in Figure 8.14. The wing length, L, is equal to 5 m, and the chord length is equal to 1 m. An isotropic material is used with $E = 75$ GPa and $\nu = 0.33$.

A torsional load is first applied via two opposite concentrated loads applied to the free-tip leading and trailing edge; the magnitude is equal to 1000 N. The results are presented in Table 8.6 and Figure 8.15. The same structural model is then investigated by means of a free-vibration analysis. Bending, torsional, and shell-like modes are described in Tables 8.7, 8.8, and Figure 8.16.

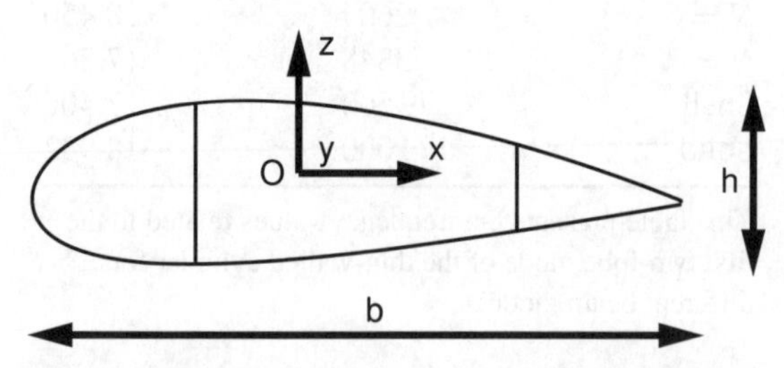

Figure 8.14 Three-cell wing model cross-section.

Table 8.6 u_z displacement at the trailing edge of the wing for different beam theories and comparison to a solid model (Carrera *et al.* 2010b).

Theory	DOFs	$u_x \times 10^5$ (m)	$u_z \times 10^3$ (m)
EBBT	155	0.0	0.0
TBT	155	0.0	0.0
$N = 1$	279	0.280	−0.074
$N = 2$	558	3.260	−0.681
$N = 3$	930	5.152	−0.818
$N = 4$	1395	5.620	−0.877
$N = 5$	1953	6.087	−0.944
$N = 6$	2604	6.477	−0.981
$N = 7$	3348	6.984	−1.029
$N = 8$	4185	7.231	−1.052
Solid	600 000	6.926	−1.305

This table shows the displacement components at the free tip trailing edge of a wing model under torsion.

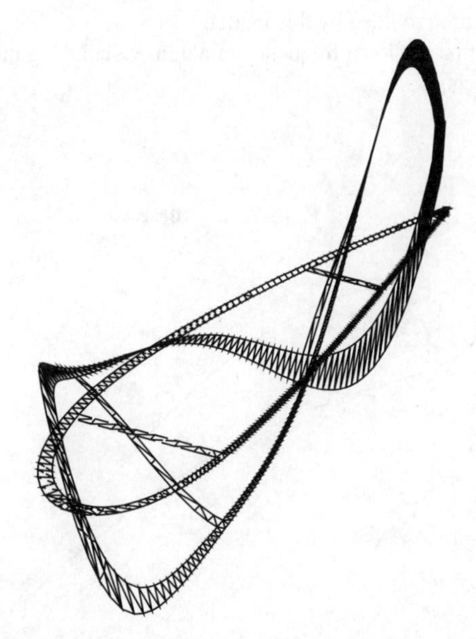

Figure 8.15 Deformed free tip of the three-cell wing model under torsion.

Table 8.7 Bending natural frequencies, in Hz, of the cantilever wing models for different theories (Carrera *et al.* 2011).

Model	EBBT	TBT	$N = 1$	$N = 2$	$N = 3$	$N = 4$	Solid
DOFs	455	455	819	1638	2730	4095	$> 6 \times 10^5$
f_1, z-dir.	5.872	5.866	5.866	5.972	5.922	5.913	5.864
f_2, x-dir.	33.340	32.709	32.709	32.834	32.656	32.625	32.335
f_3, z-dir.	36.735	36.581	36.581	37.127	36.376	36.216	34.844
f_4, z-dir.	102.634	101.617	101.617	103.210	98.918	98.133	81.976

This table shows the bending frequencies of a thin-walled wing model obtained via increasing order models.

Table 8.8 First torsional natural frequency, in Hz, of the cantilever wing models for different theories (Carrera *et al.* 2011).

Model type	EBBT	TBT	$N = 1$	$N = 2$	$N = 3$	$N = 4$	Solid
DOFs	455	455	819	1638	2730	4095	$> 6 \times 10^5$
	—*	—*	161.581	56.857	54.457	53.979	44.481

*No torsional modes are provided by this model.
This table shows the first torsional frequency of a thin-walled wing model obtained via increasing order models.

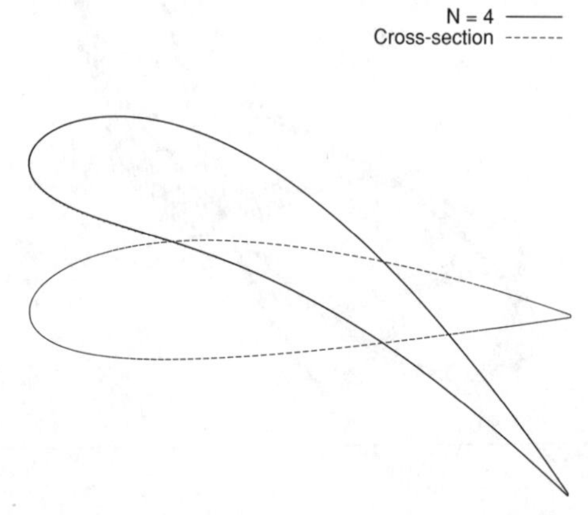

Figure 8.16 Wing cross-section, 17th natural modal shape, $f = 604$ Hz (Carrera *et al.* 2011).

Table 8.9 Bridge-like cross-section dimensions.

	m
j	15.200
k	7.300
m	3.450
n	2.155
r	1.295
L	100

This table presents the geometric characteristics of the bridge cross-section in Figure 8.17.

An analysis of the airfoil-shaped model shows the strength of the present beam formulation in dealing with thin-walled structures with arbitrary geometries. The following considerations are of particular interest:

- The use of higher-order beams permits us to detect shell-like phenomena with a significant reduction in the computational costs.
- Classical models are not able to detect torsion and shell-like results.
- Models higher than fourth order are needed to detect the torsion of a thin-walled structure.

8.4 Free vibrations of a bridge-like beam

The free vibrations of a bridge cross-section beam model (see Table 8.9, Figure 8.17) are considered in this section. The geometric characteristics of the structure are taken from the works of Gruttmann and Wagner (2001) and Gruttmann *et al.* (1999). Young's modulus, E, is equal to 210 GPa, the Poisson ratio, ν, is

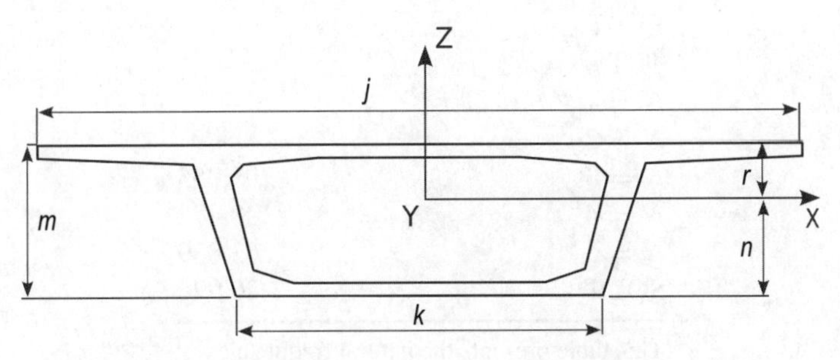

Figure 8.17 Bridge-like cross-section (Petrolo *et al.* 2011).

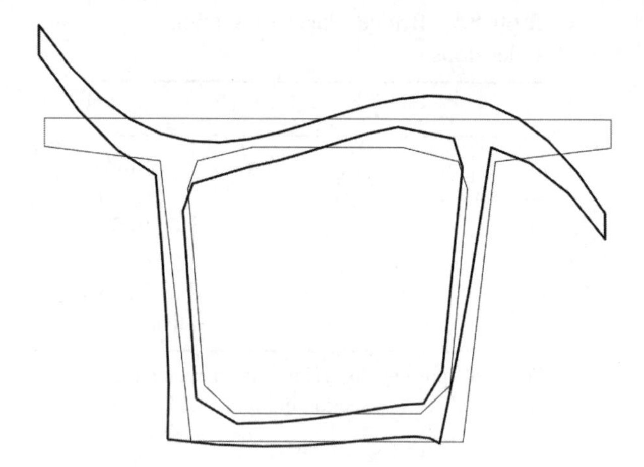

Figure 8.18 Bridge-like cross-section shell-like mode via a $N = 5$ beam model, $f = 63.839$ Hz (Petrolo *et al.* 2011).

equal to 0.33, and the density, ρ, is equal to $7900\,\text{kg/m}^3$. A 10-point B4 beam element mesh is used and the beam is clamped at both ends. Attention is given to a shell-like natural mode with an extensive in-plane distortion of the cross-section. The mid-span natural mode of the cross-section is shown in Figure 8.18. Table 8.10 presents the natural frequencies obtained via different beam models and compared to a Nastran SOLID model. The following considerations hold:

- Classical models and refined theories lower than second order are not able to detect this mode.

Table 8.10 Bridge-like cross-section natural frequencies via different models.

	Hz
EBBT	—
TBT	—
$N = 1$	—
$N = 2$	—
$N = 3$	206.055
$N = 4$	120.472
$N = 5$	63.839
SOLID	20.97

This table presents the natural frequencies related to the mode shown in Figure 8.18.

- Large differences in the natural frequency are observed as the beam order is increased, starting from the cubic model.
- The $N = 5$ beam model is still not sufficient to detect the correct natural frequency.

References

Carrera E, Giunta G, and Petrolo M 2010a A modern and compact way to formulate classical and advanced beam theories. *In: Developments in Computational Structures Technology*, Ch. 4. DOI: 10.4203/csets.25.4.

Carrera E, Giorcelli E, Mattiazzo G, and Petrolo M 2010b Refined beam models for static and dynamic analysis of wings and rotor blades. *In: Proceedings of CST2010, The Tenth International Conference on Computational Structures Technology, Valencia, Spain, 14–17 September*.

Carrera E, Petrolo M, and Varello A 2011 Advanced beam formulations for free vibration analysis of conventional and joined wings. *Journal of Aerospace Engineering*. In Press. Doi: 10.1061/(ASCE)AS.1943-5525.0000130.

Giunta G, Biscani F, Carrera E, and Belouettar S 2011 Analysis of thin-walled beams via a one-dimensional unified formulation. *International Journal of Applied Mechanics*. In Press.

Gruttmann F and Wagner W 2001 Shear correction factors in Timoshenko's beam theory for arbitrary shaped cross-sections. *Computational Mechanics*, **27**, 199–207.

Gruttmann F, Sauer R, and Wagner W 1999 Shear stresses in prismatic beams with arbitrary cross-sections. *International Journal for Numerical Methods in Engineering* **45**, 865–889.

Petrolo M, Zappino E, and Carrera E 2011 Refined free vibration analysis of one-dimensional structures with compact and bridge-like cross-sections. Submitted.

9

Linearized elastic stability

In this chapter, after a brief discussion of the Euler classical solution, the linearized elastic stability of beam structures is presented within the CUF. The proposed solutions are: (1) linearized in the sense that the change in geometry due to pre-buckling deformation is disregarded; and (2) elastic since they are valid only in the case of a critical buckling stress lower than the material elastic limit strength. It is also assumed that the beams are initially perfectly straight.

9.1 Critical buckling load classic solution

The classical formula for the calculation of the critical buckling load of a beam was developed by Euler in 1744. The problem solved by Euler was a perfectly straight, elastic, and slender beam that is clamped at one end and compressed by a centrally applied load on the other end, see Figure 9.1. The critical buckling load P_{cr} for this case is

$$P_{cr} = \pi^2 \frac{EI_{min}}{4L^2} \tag{9.1}$$

where I_{min} is the minimum moment of inertia of the cross-section. As soon as the external load is equal to P_{cr}, the straight equilibrium configuration becomes unstable and an infinitesimal perturbation makes the beam reach a new stable configuration as shown in Figure 9.2. For the problem under consideration here, the critical buckling load is proportional to the cross-section bending stiffness (which accounts for both material and geometry) and inversely proportional to

Beam Structures: Classical and Advanced Theories, First Edition. Erasmo Carrera, Gaetano Giunta and Marco Petrolo.

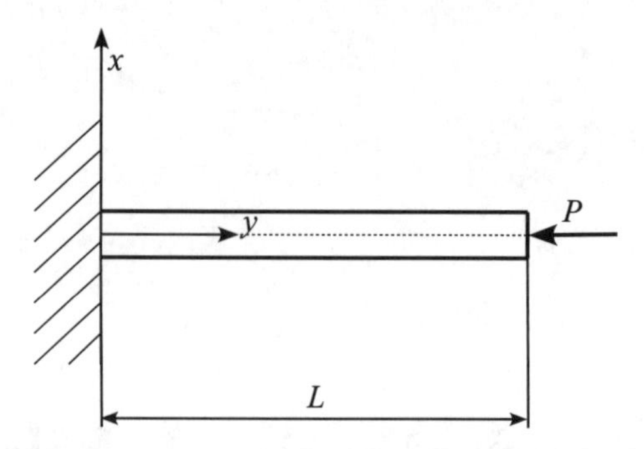

Figure 9.1 Perfectly straight and elastic slender beam compressed by a centrally applied load.

the square of the length of the beam. A general solution that accounts for several different boundary conditions is formally similar to Equation 9.1:

$$P_{cr} = \pi^2 \frac{EI_{min}}{\tilde{L}^2} \tag{9.2}$$

where $\tilde{L}$ is the distance along the beam axis between the two points at which the curvature of the buckled configuration of the beam changes as shown in Figures 9.2 and 9.3. Table 9.1 presents the value of $\tilde{L}$ for different geometric boundary conditions. The critical stress σ_{cr} is obtained by dividing the critical load by the cross-sectional area:

$$\sigma_{cr} = \frac{P_{cr}}{A} = \pi^2 \frac{EI_{min}}{A\tilde{L}^2} \tag{9.3}$$

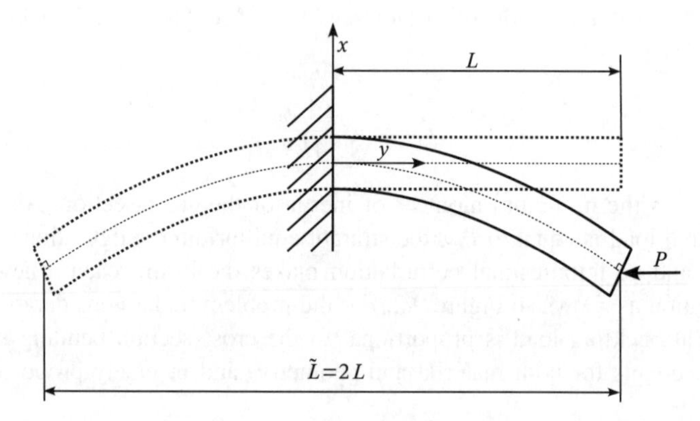

Figure 9.2 Bending buckling mode for a perfectly straight and slender clamped beam.

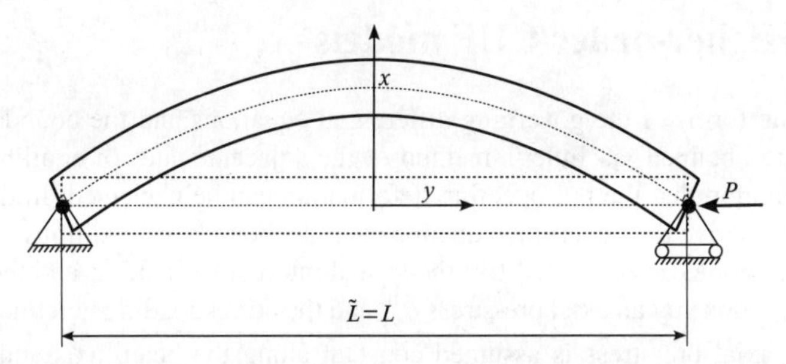

Figure 9.3 Bending buckling mode for a perfectly straight and elastic simply supported beam.

Upon introducing the radius of gyration of the cross-section

$$\rho_\Omega = \sqrt{\frac{I_{min}}{A}} \tag{9.4}$$

Equation 9.3 becomes

$$\sigma_{cr} = \frac{\pi^2 E}{\left(\tilde{L}/\rho_\Omega\right)^2} \tag{9.5}$$

As far as the stability of thin-walled beams is concerned, the classical solutions are those obtained by Wagner (1929, 1936), Goodier (1942), Timoshenko (Timoshenko, 1945, Timoshenko and Goodier, 1970), Bleich (1952), and Vlasov (1959). Governing differential equations were derived from either geometric/equilibrium considerations or the principle of stationary potential energy. For a more detailed explanation of the stability of beams, the interested reader should refer to the books by Timoshenko and Gere (1961) and by Simitses (1986).

Table 9.1 Values of $\tilde{L}$ for different boundary conditions.

Boundary conditions	$\tilde{L}$
Simply supported–simply supported	L
Clamped–free	$2L$
Clamped–clamped	$\frac{1}{2}L$
Clamped–simply supported	$\frac{\pi}{4.493}L$

This table presents the value of $\tilde{L}$ in the Euler critical buckling equation for different boundary conditions.

9.2 Higher-order CUF models

The strong forms of the governing differential equations and the boundary conditions are obtained via Euler's method of the adjacent states of equilibrium on the assumption that the pre-buckling deformation can be neglected. In the PVD, in order to derive the governing equations and the boundary conditions, two contributions should be accounted for: the virtual internal work δL_{int} and the virtual work $\delta L_{\sigma^0_{yy}}$ done by an axial pre-stress σ^0_{yy} and the corresponding non linear strain ϵ^{nl}_{yy}. The axial pre-stress is assumed constant along the beam axis and it does not change either in magnitude or in direction during buckling. The non linear strain is intended in the Green–Lagrange sense, the quadratic derivatives of the displacement components all being retained:

$$\epsilon^{nl}_{yy} = \frac{1}{2}\left(u^2_{x,y} + u^2_{y,y} + u^2_{z,y}\right) \tag{9.6}$$

In the case of linearized stability analysis, the PVD is

$$\delta L_{int} + \delta L_{\sigma^0_{yy}} = 0 \tag{9.7}$$

The virtual variation of the strain energy has been discussed in Chapter 5. The virtual work of the axial pre-stress is

$$\delta L_{\sigma^0_{yy}} = \int_L \int_\Omega \sigma^0_{yy} \delta\epsilon^{nl}_{yy} \, d\Omega dy \tag{9.8}$$

Upon substitution of Equations 9.6 and the approximation of the displacement vector in CUF form, and after integration by parts, Equation 9.8 becomes

$$\delta L_{\sigma^0_{yy}} = -\sigma^0_{yy} \int_L \delta \mathbf{u}^T_\tau \int_\Omega F_\tau F_s d\Omega \, \mathbf{I} \frac{\partial^2}{\partial y^2} \mathbf{u}_s \, dy + \sigma^0_{yy} \, \delta \mathbf{u}^T_\tau \int_\Omega F_\tau F_s d\Omega \, \mathbf{I} \frac{\partial^2}{\partial y^2} \mathbf{u}_s \Big|_{y=0}^{y=L} \tag{9.9}$$

This equation can be put into a compact vectorial form:

$$\delta L_{\sigma^0_{yy}} = -\sigma^0_{yy} \int_L \delta \mathbf{u}^T_\tau \, \overline{\mathbf{K}}^{\tau s}_{\sigma^0_{yy}} \, \mathbf{u}_s \, dy + \delta \mathbf{u}^T_\tau \, \Pi^{\tau s}_{\sigma^0_{yy}} \, \mathbf{u}_s \Big|_{y=0}^{y=L} \tag{9.10}$$

$\overline{\mathbf{K}}^{\tau s}_{\sigma^0_{yy}}$ is the differential geometric stiffness matrix. Its components are

$$\overline{K}^{\tau s}_{ij\sigma^0_{yy}} = \delta_{ij} J_{\tau s} \frac{\partial^2}{\partial y^2} \mathbf{I} \text{ with } i, j = x, y, z \tag{9.11}$$

where δ_{ij} is Kronecker's delta and

$$J_{\tau s} = \int_{\Omega} F_\tau F_s \, d\Omega \tag{9.12}$$

$\Pi^{\tau s}_{\sigma^0_{yy}}$ is the differential matrix of the boundary conditions:

$$\Pi^{\tau s}_{ij\sigma^0_{yy}} = \delta_{ij} J_{\tau s} \frac{\partial}{\partial y} \mathbf{I} \text{ with } i, j = x, y, z \tag{9.13}$$

9.2.1 Governing equations, fundamental nucleus

The explicit form of the fundamental nucleus of the governing equations is

$$\begin{aligned}
\delta u_{x\tau} &: \left(J^{22}_{\tau_{,x}s_{,x}} + J^{66}_{\tau_{,z}s_{,z}} + \alpha^2 J^{44}_{\tau s}\right) u_{xs} - \alpha \left(J^{23}_{\tau_{,x}s} - J^{44}_{\tau s_{,x}}\right) u_{ys} \\
&\quad + \left(J^{12}_{\tau_{,x}s_{,z}} + J^{66}_{\tau_{,z}s_{,x}}\right) u_{zs} - \sigma^0_{yy} J_{\tau s} u_{xs,_{yy}} = 0 \\
\delta u_{y\tau} &: -\alpha \left(J^{23}_{\tau s_{,x}} - J^{44}_{\tau_{,x}s}\right) u_{xs} + \left(J^{44}_{\tau_{,x}s_{,x}} + J^{55}_{\tau_{,z}s_{,z}} + \alpha^2 J^{33}_{\tau s}\right) u_{ys} \\
&\quad -\alpha \left(J^{13}_{\tau s_{,z}} - J^{55}_{\tau_{,z}s}\right) u_{zs} - \sigma^0_{yy} J_{\tau s} u_{ys,_{yy}} = 0 \\
\delta u_{z\tau} &: \left(J^{12}_{\tau_{,z}s_{,x}} + J^{66}_{\tau_{,x}s_{,z}}\right) u_{xs} - \alpha \left(J^{13}_{\tau_{,z}s} - J^{55}_{\tau s_{,z}}\right) u_{ys} \\
&\quad + \left(J^{11}_{\tau_{,z}s_{,z}} + J^{66}_{\tau_{,x}s_{,x}} + \alpha^2 J^{55}_{\tau s}\right) u_{zs} - \sigma^0_{yy} J_{\tau s} u_{zs,_{yy}} = 0
\end{aligned} \tag{9.14}$$

and the boundary conditions are

$$\begin{aligned}
&\delta u_{x\tau} \left(J^{44}_{\tau s} u_{xs,_y} + J^{44}_{\tau s_{,x}} u_{ys} + \sigma^0_{yy} J_{\tau s} u_{xs,_y}\right)\Big|_{y=0}^{y=L} = 0 \\
&\delta u_{y\tau} \left(J^{13}_{\tau s_{,z}} u_{zs} + J^{23}_{\tau s_{,x}} u_{xs} + J^{33}_{\tau s} u_{ys,_y} + \sigma^0_{yy} J_{\tau s} u_{ys,_y}\right)\Big|_{y=0}^{y=L} = 0 \\
&\delta u_{z\tau} \left(J^{55}_{\tau s} u_{zs,_y} + J^{55}_{\tau s_{,z}} u_{ys} + \sigma^0_{yy} J_{\tau s} u_{zs,_y}\right)\Big|_{y=0}^{y=L} = 0
\end{aligned} \tag{9.15}$$

For a fixed approximation order, the nucleus has to be expanded versus the indexes τ and s in order to obtain the governing equations and the boundary conditions of the desired model.

9.2.2 Closed form analytical solution

The fundamental nucleus of the algebraic eigensystem is obtained by substituting the displacement field typical of the Navier-type closed form solution (see Equations 5.25) into Equations 9.15:

$$\delta \mathbf{U}_\tau : \left(\mathbf{K}^{\tau s} + \sigma^0_{yy} \mathbf{K}^{\tau s}_{\sigma^0_{yy}}\right) \mathbf{U}_s = 0 \tag{9.16}$$

The components of the geometrical stiffness matrix $\mathbf{K}^{\tau s}_{\sigma^0_{yy}}$ are

$$K^{\tau s}_{ij\sigma^0_{yy}} = \delta_{ij}\alpha^2 J_{\tau s}\mathbf{I} \text{ with } i, j = x, y, z \tag{9.17}$$

The solution yields as many eigenvalues (or buckling stresses) and eigenvectors (or buckling modes) as the degrees of freedom of the model.

9.3 Examples

Two examples are presented. In the first one, the critical buckling load of a beam is addressed. Higher-order models and Euler formula are compared. The minimum value of the slenderness ratio for the buckling to occur according to an elastic material behavior is discussed. In the second example, a thin plate is investigated in order to highlight the shell capabilities of the proposed models.

Example 9.3.1 *A simply supported beam with a square cross-section is considered. The beam is made of aluminum alloy 7075-T6 whose mechanical properties are: Young's modulus* 71 700 MPa, *Poisson ratio* 0.3, *and linear limit stress* ($\sigma_{p0.2}$) 503 MPa.

The minimum slenderness ratio $(L/a)_{min}$ *for a critical buckling stress within the linear elastic behavior of the material is determined first. The stress–strain relation for the material is linear as long as the stress level is lower than or equal to the linear limit stress* $\sigma_{p0.2}$. *The minimum slenderness ratio can be obtained by imposing that the critical buckling load should be lower than or equal to the linear limit stress:*

$$\sigma^0_{yy,cr} \leq \sigma_{p0.2} \tag{9.18}$$

The Euler formula for a square cross-section is

$$\sigma^0_{yy,cr} = \pi^2 \frac{E\tilde{I}}{A\tilde{L}} = \frac{\pi^2}{12} E \left(\frac{a}{L}\right)^2 \tag{9.19}$$

The minimum slenderness ratio is obtained upon substituting Equation 9.19 into Equation 9.18:

$$(L/a)_{min} = \pi \sqrt{\frac{E}{12\sigma_{p0.2}}} \tag{9.20}$$

Table 9.2 Buckling loadings.

L/a	100	50	20	15
Euler	5.897	23.588	147.427	262.093
$N = 5, 6$	5.895	23.558	146.232	258.344
$N = 4$	5.895	23.558	146.233	258.347
$N = 3$	5.895	23.558	146.233	258.348
$N = 2$	5.895	23.561	146.360	258.742
$N = 1$	5.895	23.560	146.346	258.697

Bending buckling loadings (in MPa) for a square cross-section beam.

It should be noted that the minimum slenderness ratio depends upon the material properties only. For the aluminum alloy 7075-T6 the previous equation yields

$$(L/a)_{min} = \pi\sqrt{\frac{71\,700}{12\,503}}\ \text{MPa} = 10.83\ \text{MPa} \tag{9.21}$$

For slenderness ratios lower than $(L/a)_{min}$, Young's modulus is no longer constant versus the stress. It can be proved that the Euler equation can be still formally used in the case of inelastic buckling upon substitution of the constant Young's modulus with another one reduced accordingly. The critical buckling loadings for a slenderness ration as low as 15 are presented in Table 9.2. Results are obtained by means of the Euler formula and higher-order beam models. The results converge for an expansion order as high as five. The difference between the Euler formula and higher-order theories increases with decreasing slenderness ratio. The flexural, torsional, and axial buckling modes for m *equal to one and three are presented in Figures 9.4 to 9.6.*

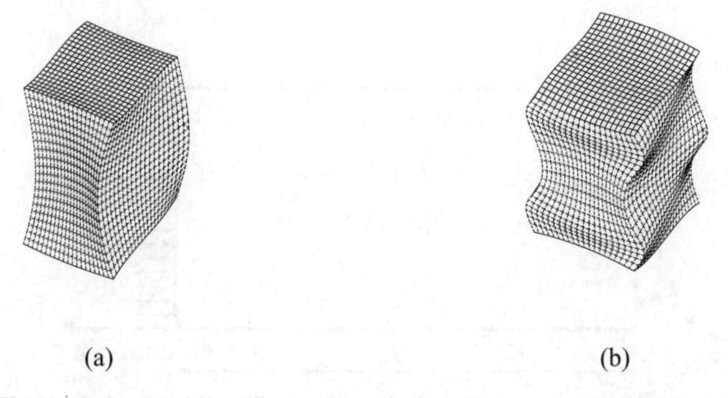

(a) (b)

Figure 9.4 Buckling flexural mode for (a) $m = 1$ and (b) $m = 3$.

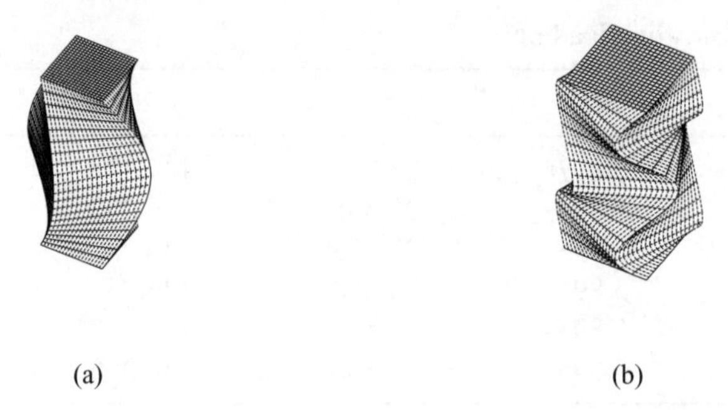

Figure 9.5 Buckling torsional mode for (a) $m = 1$ and (b) $m = 3$.

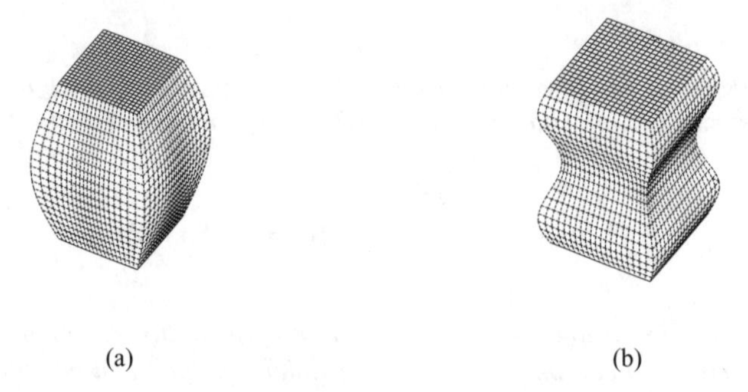

Figure 9.6 Buckling axial mode for (a) $m = 1$ and (b) $m = 3$.

Example 9.3.2 *In order to highlight the shell capabilities of the developed models, a thin simply supported plate is considered. The plate is made of the same material as in the previous example. The plate's geometry is such that the length L is equal to* 0.1 m, *the width is* 0.025 m *and it is* 1 mm *thick. Geometry, loading, and constraints are reported in Figure 9.7. The first four bending and*

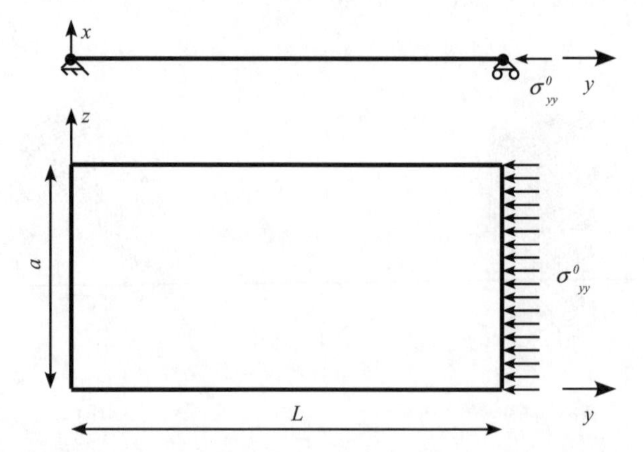

Figure 9.7 Plate geometry, loading, and boundary conditions.

Table 9.3 Bending and twisting buckling loadings.

	Bending				Twisting			
m	1	2	3	4	1	2	3	4
FEM	5.932	24.03	54.76	98.25	181.61	198.78	229.16	272.31
$N = 7$	5.932	24.03	54.71	98.08	180.38	198.88	229.68	272.72
$N = 5$	5.933	24.04	54.74	98.14	181.28	199.50	230.02	272.92
$N = 3$	5.934	24.08	54.98	98.85	181.18	201.09	232.74	277.00
$N = 1$	5.895	23.56	52.93	93.91	3125.5	8803.8	13611	17087

First four bending and twisting buckling loadings (in MPa) for a thin simply supported plate.

twisting buckling loads are computed and compared to a FEM reference solution. This reference solution is obtained via the commercial code Nastran. The four-node, two-dimensional "QUAD4" element is used, see Getting Started with MSC. Nastran, User's Guide (2001). The side length of each element is 0.5 mm*. The total number of degrees of freedom is about* 51×10^3*, whereas in the case of a seventh-order analytical model, it is* 108*. Results are presented in Table 9.3.*

They are all below the yielding stress, therefore a liner stress–strain relation can be assumed. Accurate results are obtained for any value of m *when the expansion order is as low as three, being the maximum difference from the FEM solution of about* 0.6% *and* 1.7% *for the bending and twisting modes, respectively. A first-order model is accurate for bending modes with* m *as high as two. Twisting buckling loadings are greatly overestimated since the cross-section is too stiff. The flexural and twisting buckling shapes for* m *equal to one and four are reported in Figures 9.8 and 9.9, respectively.*

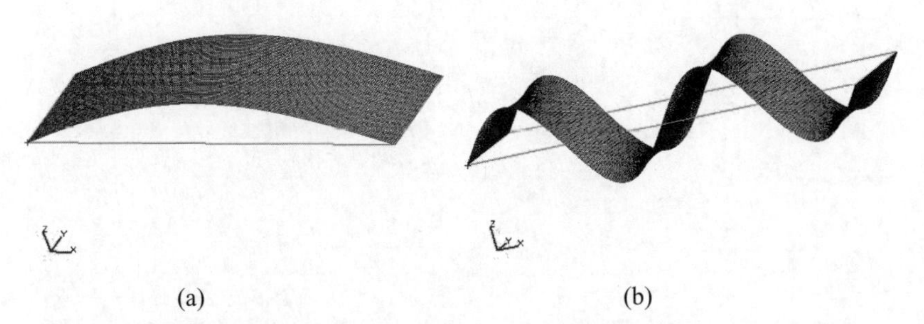

Figure 9.8 Bending buckling shapes for m equal to (a) one and (b) four.

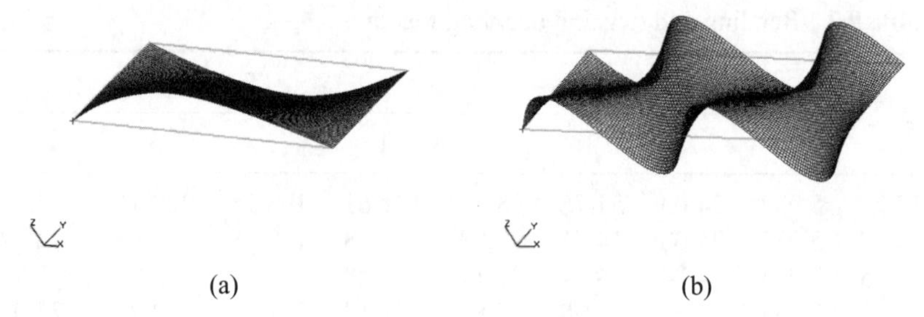

(a) (b)

Figure 9.9 Twisting buckling shapes for m equal to (a) one and (b) four.

References

Bleich F 1952 *Buckling Strength of Metal Structures*. McGraw-Hill.

Getting Started with MSC.Nastran User's Guide. MSC.Software Corporation, Santa Ana, CA.

Goodier JN 1942 Flexural-torsional buckling of bars of open section under bending, eccentric thrust or torsional loads. *Cornell University Engineering Experimental Station Bulletin*, **28**.

Simitses GJ 1986 *An Introduction to the Elastic Stability of Structures*. Prentice Hall.

Timoshenko SP 1945 Theory of bending, torsion and buckling of thin-walled members of open cross section. *Journal of the Franklin Institute*, **239**(3), 201–219.

Timoshenko SP and Gere JM 1961 *Theory of Elastic Stability*. Dover.

Timoshenko SP and Goodier JN 1970 *Theory of elasticity*. McGraw-Hill.

Vlasov VZ 1959 *Thin-walled elastic beams*. Fizmatgiz.

Wagner H 1929 *Verdrehung und Knickung von offenen Profilen*. The 25th Anniversary of the Technische Hochschule, Danzig, 1929.

Wagner H 1936 *Torsion and buckling of open sections*. Technical Memorandum No. 807, National Advisory Committee for Aeronautics, USA, 1936.

10

Beams made of functionally graded materials

Functionally graded materials (FGMs) are briefly presented, accounting for their advantages when compared to conventional materials, their applications, and some theories for predicting the mechanical properties and structural behavior of FGM-based structures. Some gradation laws for isotropic FGMs are then addressed. Finally, it is shown how to account for material gradation in the modeling of beam structures within the CUF, and, finally, two examples are discussed.

10.1 Functionally graded materials

FGMs are composed of two or more materials whose volume fraction changes gradually along desired spatial directions resulting in a smooth and continuous change in the effective properties. The combination of different materials with specific physical properties allows a tailored material design that broadens the structural design space by implementing a multi-functional response with a minimal weight increase. Figure 10.1 presents a FGM structure in which the material changes gradually from full metallic at the bottom to full ceramic at the top. The metallic area is used to withstand the mechanical loads, while the ceramic one acts as thermal protection.

The concept of FGM was developed by Japanese material scientists in 1984 for super heat-resistant materials to be used in spaceplanes. The first feasibility study, "The basic technology for the development of functionally gradient materials for

Beam Structures: Classical and Advanced Theories, First Edition. Erasmo Carrera, Gaetano Giunta and Marco Petrolo.

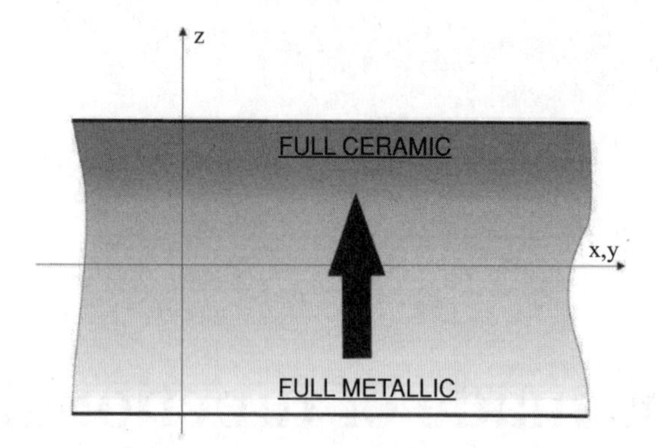

Figure 10.1 An example of FGM structure.

relaxation of thermal stresses" was carried out 1986, see Koizumi (1997). At the end of the study, a square shell of SiC–C FGM for the base of the fuselage of spaceplanes was fabricated.

The gradual and smooth change in the material profile and in the effective physical properties achievable by FGMs distinguishes them from conventional materials since:

- the mismatch that is often encountered in laminated composites and is known to promote delamination-related problems can be eliminated;
- in-plane and transverse through-the-thickness stresses can be reduced;
- the residual stress distribution can be improved;
- the fracture toughness can be enhanced;
- the stress intensity factors can be reduced; and
- the fatigue life can be increased.

FGMs have found many applications in different fields such as:

- energy conversion
- dental and orthopedic implants
- sensors and thermo-generators
- wear-resistant coatings
- aerospace.

For a general account of FGMs in terms of design, fabrication, and applications, the interested reader should refer to the books by Suresh and Mortensen (1998) and Miyamoto *et al.* (1999) or the review articles by Watanabe *et al.* (2003) and Birman and Byrd (2007).

Several approaches have been proposed in micro-mechanics to determine the mechanical properties of FGMs starting from the gradation law of the constituent materials. A review of classical approaches such as self-consistent schemes, differential schemes, the method by Mori and Tanaka (1973), and concentric cylinder models can be found in Aboudi (1991), Nemat-Nasser and Hori (1993), and Zuiker and Dvorak (1994). One of the first micro-mechanical approaches developed especially for FGMs was proposed by Wakashima and Tsukamoto (1990). Aboudi *et al.* (1994, 1996, 1999) introduced a higher-order theory for determining the mechanical properties of FGMs based on a volumetric averaging approach that explicitly couples micro- and macro-scales with spatially varying micro-structures in one, two, and three orthogonal directions.

As far as the modeling of FGM-based structures is concerned, Chakraborty *et al.* (2003) developed a finite element based on Timoshenko's beam theory in which the shape functions have been derived from the general exact solution of the static governing equations. Exponential and polynomial gradation of mechanical and thermal properties along the through-the-thickness direction were considered. Li (2008) proposed a unified approach to the formulation of the Timoshenko and Euler–Bernoulli beam models. Young's modulus varied along the transverse coordinate according to a power-law function. Kadoli *et al.* (2008) proposed a fined element based on a third-order approximation of the axial displacement and constant transverse displacement for the static analysis of beams made of metal–ceramic FGMs. The components' volume fraction was supposed to vary according to a power-law function. A discrete layer approach was adopted to account for material gradation. FGMs have been considered within the CUF for higher-order beam modeling by Giunta *et al.* (2010). As far as elasticity solutions are concerned, Sankar (2001) solved the plane elasticity equations exactly. An EBBT-type theory was also derived. The shear stress component was obtained via integration of the indefinite equilibrium equations. Simply supported FGM beams subjected to sinusoidal loadings were investigated. Young's modulus was supposed to vary exponentially along the thickness direction and the Poisson ratio was constant. Zhu and Sankar (2004) adopted a combined Fourier series–Galerkin method for the analysis of simply supported FGM-based beams for which Young's modulus was a third-order polynomial function of the through-the-thickness coordinate. The Poisson ratio was constant. The basic functions were the elements of the classical polynomial base up to third order versus the thickness direction. Under the hypothesis of plane stress, Ding *et al.* (2007) generalized Silverman's method (see Silverman 1964) in order to obtain a stress function for anisotropic functionally graded beams in a general manner. The governing equations were derived independently from the variation of the elastic compliance parameters along the beam thickness. Cantilever, simply supported, and fixed–fixed beams were studied. An exponential function was assumed for the material compliance coefficient S_{11}, while the others were considered constant.

10.2 Material gradation laws

Several models have been proposed in order to determine the mechanical properties of an isotropic FGM on the basis of a given gradation law. The exponential and the power gradation law are presented here.

10.2.1 Exponential gradation law

In the case of the exponential gradation law, Young's modulus is supposed to vary versus the in-plane coordinates x and z according to the following exponential law:

$$E(x, z) = E_0 e^{(\alpha_1 x + \beta_1)} e^{(\alpha_2 z + \beta_2)} \tag{10.1}$$

Coefficients α_i and β_i with $i = 1$ and 2 are additional degrees of freedom. For instance, they account for a generic reference system with respect to the gradation law. Figure 10.2 presents the variation of E/E_0 versus x and z over a rectangular cross-section. The variation of the Poisson ratio is negligible and therefore it is assumed to be constant.

10.2.2 Power gradation law

The effective properties of ceramic–metal FGMs can be described via a power gradation law for the volume fraction ϕ of the constituent materials:

$$\begin{aligned} &\phi_c(x, z) = (\alpha_1 x + \beta_1)^{n_1} (\alpha_2 z + \beta_2)^{n_2} \quad \text{with } \alpha_i,\ \beta_i,\ n_i : \phi_c \in [0, 1] \\ &\phi_m(x, z) = 1 - \phi_c(y, z) \end{aligned} \tag{10.2}$$

Subscripts "c" and "m" stand for ceramic and metallic material, respectively. Coefficients α_i and β_i have the same meaning as in Equation 10.1, n_1 and n_2 are the volume fraction exponents such that $0 \leq n_1, n_2 < \infty$. The case of $n_1 = n_2 = 0$

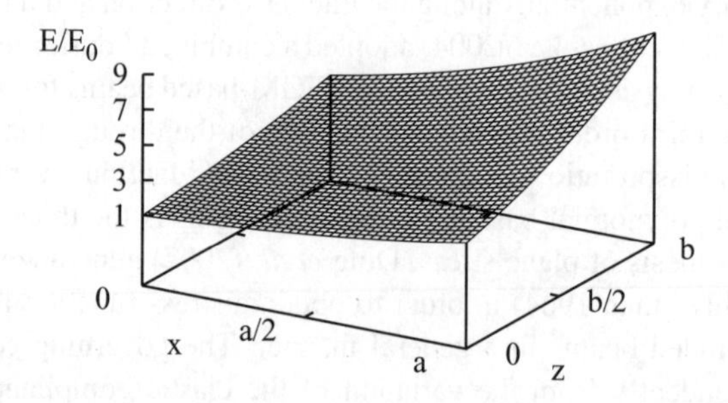

Figure 10.2 Young's modulus variation along the cross-section according to the exponential gradation law.

represents a fully ceramic material. On the contrary, a fully metallic material is obtained for α_i and β_i equal to zero. As a first approximation, the effective properties can be predicted starting from Equations 10.2 via several analytical methods such as the rule of mixture, the self-consistent method by Hill (1965), and the scheme by Mori and Tanaka (1973).

10.2.2.1 The rule of mixtures

A generic material property $f(x,z)$ can be derived from the rule of mixtures starting from the properties f_c and f_m of the constituent materials:

$$\begin{aligned} f(x,z) &= f_c\, \phi_c(x,z) + f_m\, \phi_m(x,z) \\ &= (f_c - f_m)\, \phi_c(x,z) + f_m \end{aligned} \tag{10.3}$$

The main advantage of Equation 10.3 is that it is easy to use. Figure 10.3 presents the variation of Young's modulus along the cross-section according to the rule of mixture and power-law variation for a square cross-section beam with sides of length a. The linear and the quadratic variation versus x are considered.

10.2.2.2 The modified rule of mixtures

The modified rule of mixtures was proposed by Tomota *et al.* (1976) for cemented carbides. It has been adopted by several researchers for predicting Young's

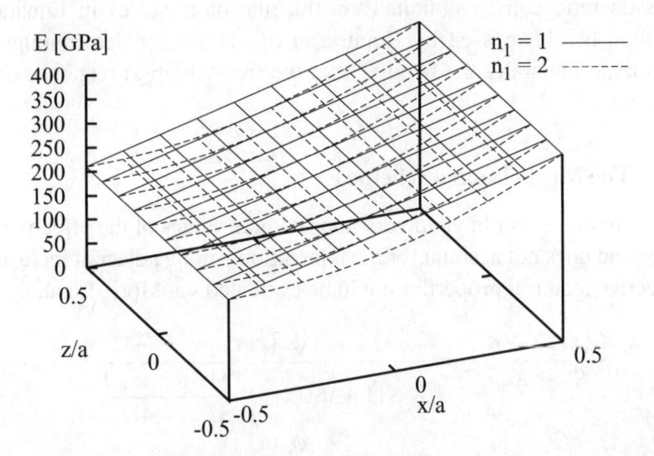

Figure 10.3 Young's modulus variation along the cross-section according to the rule of mixture and power-law variation.

modulus of two-constituent FGMs. A stepwise approach is used rather than a continuous variation, such as the one in Equation 10.2:

$$\phi_c(z) = \begin{cases} \phi_{c1} & z_{11} \leq z < z_{12} \\ \phi_{c2} & z_{12} \leq z < z_{22} \\ \vdots & \\ \phi_{cN} & z_{N-1N} \leq z < z_{NN} \end{cases} \tag{10.4}$$

where ϕ_{ci} are constant values, N is the total number of layers in with the FGM is divided, and z_{ij} are the bottom and top coordinates of each layer. Each layer is considered to be isotropic. The uni-axial stress (σ) and strain (ϵ) in each layer are related to the average uniaxial stresses σ_m, σ_c and strains ϵ_m and ϵ_c of the two phases:

$$\begin{aligned} \sigma &= \sigma_c \phi_c + \sigma_m \phi_m \\ \epsilon &= \epsilon_c \phi_c + \epsilon_m \phi_m \\ q &= -\frac{\sigma_c - \sigma_m}{\epsilon_c - \epsilon_m} \end{aligned} \tag{10.5}$$

where q is the ratio of stress to strain transfer between the two phases. This value has been determined experimentally by Finot *et al.* (1996) for a Ni/Al_2O_3 FGM ($q = 4.5\,\text{GPa}$) and by Bhattacharyya *et al.* (2007) for an Al/SiC FGM ($q = 91.6\,\text{GPa}$). The effective elastic modulus is

$$E = \frac{\phi_m E_m (q + E_c) + E_c (1 - \phi_m)(q + E_m)}{\phi_m (q + E_c) + (1 - \phi_m)(q + E_m)} \tag{10.6}$$

The Poisson ratio can be obtained via the rule of mixtures in Equation 10.3. This method has been used by Kapuria *et al.* (2008) for determining numerically and experimentally the bending and the free-vibration response of layered FGM beams.

10.2.2.3 The Mori–Tanaka scheme

The rule of mixtures could yield very approximate values of the effective material properties and does not account for the interaction among adjacent inclusions. The local effective material properties could be evaluated via Mori–Tanaka's scheme

$$\begin{aligned} \frac{K(x,z) - K_m}{K_c - K_m} &= \frac{\phi_c(x,z)}{1 + [1 - \phi_c(x,z)] \dfrac{3(K_c - K_m)}{3K_m + 4G_m}} \\ \frac{G(x,z) - G_m}{G_c - G_m} &= \frac{\phi_c(x,z)}{1 + [1 - \phi_c(x,z)] \dfrac{G_c - G_m}{G_m + \dfrac{G_m(9K_m + 8G_m)}{6(K_m + 2G_m)}}} \end{aligned} \tag{10.7}$$

where K is the bulk modulus. The effective Young modulus and the Poisson ratio are

$$E(x,z) = \frac{9K(x,z)\,G(x,z)}{3K(x,z)+G(x,z)}$$
$$\nu(y,z) = \frac{3K(x,z)-2G(x,z)}{2\left[3K(x,z)+G(x,z)\right]} \tag{10.8}$$

10.2.2.4 Hill's self-consistent method

Hill's self-consistent method is now presented assuming that the inclusions are spheres distributed in such a manner that the composite is statistically isotropic. The effective bulk and shear moduli are given implicitly:

$$K(x,z) = \frac{1}{\dfrac{\phi_m(x,z)}{K_m+\frac{4}{3}G(x,z)} + \dfrac{\phi_c(x,z)}{K_c+\frac{4}{3}G(x,z)}} - \frac{4}{3}G(x,z)$$
$$\left[\frac{\phi_m(x,z)\,K_m}{K_m+\frac{4}{3}G(x,z)} + \frac{\phi_c(x,z)\,K_c}{K_c+\frac{4}{3}G(x,z)}\right] + 5\left[\frac{\phi_m(x,z)\,G_c}{G(x,z)-G_c} + \frac{\phi_c(x,z)\,G_m}{G(x,z)-G_m}\right] + 2 = 0 \tag{10.9}$$

The effective shear modulus must be first determined by solving the second equation in Equation 10.9. Hill demonstrated that, within the limits of application of the method, there exists precisely one positive root that lies between the Reuss and Voigt estimates. He also provided an explicit solution in the case of diluted dispersion. Young modulus and the Poisson ratio are obtained through Equations 10.8.

10.3 Beam modeling

At a structural level (macro-scale), the gradation of the material is accounted for in the material constitutive equations. Under the hypothesis of isotropic linear elastic FGMs, the generalized Hooke's law holds:

$$\boldsymbol{\sigma}_p = \mathbf{C}_{pp}(x,z)\,\boldsymbol{\epsilon}_p + \mathbf{C}_{pn}(x,z)\,\boldsymbol{\epsilon}_n,$$
$$\boldsymbol{\sigma}_n = \mathbf{C}_{np}(x,z)\,\boldsymbol{\epsilon}_p + \mathbf{C}_{nn}(x,z)\,\boldsymbol{\epsilon}_n \tag{10.10}$$

where now matrices $\mathbf{C}_{pp}$, $\mathbf{C}_{nn}$, $\mathbf{C}_{pn}$, and $\mathbf{C}_{np}$ are also function of the in-plane coordinates and they have to be integrated over the cross-section when computing the cross-section moments $J^{ij}_{\tau(,\varphi)s(,\xi)}$ in the nucleus:

$$J^{ij}_{\tau(,\varphi)s(,\xi)} = \int_{\Omega} C_{ij}\left(x, z\right) F_{\tau(,\varphi)} F_{s(,\xi)}\, d\Omega \tag{10.11}$$

The function to be integrated, due to its physical meaning and for the gradation laws, is normally smooth enough that a numerical solution can be obtained via Gauss–Legendre quadrature:

$$\begin{aligned} J^{ij}_{\tau(,\varphi)s(,\xi)} &= \int_{-1}^{1}\int_{-1}^{1} C_{ij}\left(\overline{x}, \overline{z}\right) F_{\tau(,\varphi)}\left(\overline{x}, \overline{z}\right) F_{s(,\xi)}\left(\overline{x}, \overline{z}\right) \left|J_{\overline{xz}}\right|\, d\overline{x}d\overline{z} \\ &\approx \left|J_{\overline{xz}}\right| \sum_{h=1}^{n}\sum_{k=1}^{n} C_{ij}\left(\overline{x}_h, \overline{z}_k\right) F_{\tau(,\varphi)}\left(\overline{x}_h, \overline{z}_k\right) F_{s(,\xi)}\left(\overline{x}_h, \overline{z}_k\right) w_h w_k \end{aligned} \tag{10.12}$$

where $\left|J_{\overline{xz}}\right|$ is the determinant of the Jacobian matrix for the transformation of the coordinates x and z to the coordinates $\overline{x}$ and $\overline{z}$ such that $-1 \leq \overline{x}, \overline{z} \leq 1$, w_h and w_k are the weighting coefficients, $\overline{x}_h$ and $\overline{z}_k$ are the coordinates of the integration points, and n is the total number of Gauss points for each axis. In the case when Young's modulus varies according to the exponential law in Equation 10.1, the Poisson ratio is constant, and with rectangular cross-sections ($x_1 \leq x \leq x_2$, $z_1 \leq z \leq z_2$), an analytical solution of the integral in Equation 10.11 can be obtained. The product of the base functions or their first derivatives can be written without any loss in generality as

$$F_{\tau(,\varphi)} F_{s(,\xi)} = k_x k_z x^{n_x} z^{n_z} \tag{10.13}$$

where k_x and k_z are constant and due to differentiation, if any, or are equal to one (in the case of no derivation versus x or z). $C_{ij}\left(\overline{x}, \overline{z}\right)$ can be split as the product of a constant part and a variable one:

$$C_{ij}\left(\overline{x}, \overline{z}\right) = \overline{C}_{ij} e^{(\alpha_1 x)} e^{(\alpha_2 z)} \tag{10.14}$$

where

$$\begin{aligned} \overline{C}_{11} = \overline{C}_{22} = \overline{C}_{33} &= \frac{1-\nu}{(1+\nu)(1-2\nu)} E_0 e^{(\beta_1+\beta_2)} \\ \overline{C}_{12} = \overline{C}_{13} = \overline{C}_{23} &= \frac{\nu}{(1+\nu)(1-2\nu)} E_0 e^{(\beta_1+\beta_2)} \\ \overline{C}_{44} = \overline{C}_{55} = \overline{C}_{66} &= \frac{1}{2(1+\nu)} E_0 e^{(\beta_1+\beta_2)} \end{aligned} \tag{10.15}$$

Equation 10.11 reads

$$J^{ij}_{\tau(,_\varphi)s(,_\xi)} = \overline{C}_{ij}k_xk_z \int_{x_1}^{x_2} e^{(\alpha_1 x)}x^{n_x}\, dx \int_{z_1}^{z_2} e^{(\alpha_2 z)}z^{n_z}\, dz \tag{10.16}$$

and its analytical solution is:

$$J^{ij}_{\tau(,_\varphi)s(,_\xi)} = \overline{C}_{ij}k_xk_z \sum_{\lambda=0}^{n_x}(-1)^{n_x-\lambda}\frac{n_x!}{\lambda!}\frac{e^{(\alpha_1 x)}x^{\lambda}}{\alpha_1^{n_x+1-\lambda}}\Bigg|_{x_1}^{x_2} \sum_{\zeta=0}^{n_z}(-1)^{n_z-\zeta}\frac{n_z!}{\zeta!}\frac{e^{(\alpha_2 z)}z^{\zeta}}{\alpha_2^{n_z+1-\zeta}}\Bigg|_{z_1}^{z_2} \tag{10.17}$$

This integral can be used in the case for any beam cross-section that is obtainable as a union of rectangles. It can be demonstrated that when α_1 and α_2 tend to zero, the case of no material gradation can be obtained:

$$\lim_{\alpha_1,\alpha_2\to 0} J^{ij}_{\tau(,_\varphi)s(,_\xi)} = \overline{C}_{ij}\frac{k_x}{n_x+1}\left[(x_2)^{n_x+1}-(x_1)^{n_x+1}\right]\frac{k_z}{n_z+1}\left[(z_2)^{n_z+1}-(z_1)^{n_z+1}\right] \tag{10.18}$$

As a concluding remark, it should be noticed that since the material gradation is considered to take place only on the cross-section, the previous considerations hold for both strong and weak solutions.

10.4 Examples

Two numerical examples are presented. The Navier-type closed form solution is used. The results are assessed versus reference solutions present in the literature or three-dimensional FEM models. Analyses cover both slender and deep beams. Young's modulus is considered to vary exponentially with respect to z or both y and z coordinates. The Poisson ratio is constant.

Example 10.4.1 *A beam with a rectangular cross-section is considered. The reference system is such that* $0 \le x \le a$ *and* $-b/2 \le z \le b/2$ *with* a *equal to* 0.1 m *and the cross-section aspect ratio* a/b *equal to* 100, *see Figure 10.4. The dimension along the z-axis is negligible and the derivatives versus this coordinate can therefore be disregarded. The problem is bidimensional. Only displacements* u_x *and* u_y *and stresses* σ_{xx}, σ_{yy}, *and* σ_{xy} *are different from zero. The span-to-height ratio* L/a *is as high as* 100 *(slender beam) and as low as* 5 *(deep beam). The beam undergoes a sinusoidal pressure of maximal amplitude* $P_{xx} = 1$ Pa *and* $m = 1$ *applied on the surface at* $x = a$. *Young's modulus is graded according to the*

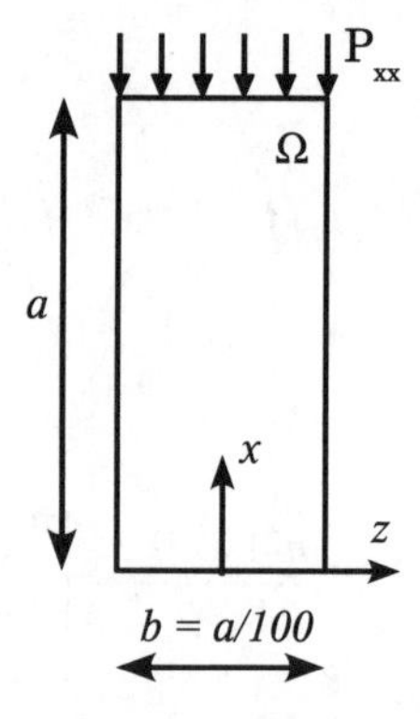

Figure 10.4 Cross-section geometry, reference system, and loading for the 2D problem.

exponential law, see Equation 10.1, with $E_0 = 1000\,\text{MPa}$, $\beta_1 = \beta_2 = 0$, $\alpha_2 = 0$, *and* $E\,(x = a)\,/E_0 = 10$. *The Poisson ratio is equal to* 0.25. *The following results are considered:*

$$
\begin{aligned}
\overline{u}_x &= u_x\,(x = 0,\, y = L/2) \\
\overline{\sigma}_{yy} &= \sigma_{yy}\,(x = 0,\, y = L/2) \\
\overline{\sigma}^a_{xx} = \sigma_{xx}\,(x = a,\, y = L/2) \quad &\overline{\sigma}^b_{xx} = \sigma_{xx}\,(x = 0,\, y = L/2) \\
\overline{\sigma}^a_{xy} = \sigma_{xy}\,(x = a/2,\, y = 0) \quad &\overline{\sigma}^b_{xy} = \sigma_{xy}\,(x = a,\, y = 0)
\end{aligned}
\tag{10.19}
$$

The values of the stresses $\overline{\sigma}^b_{xy}$, $\overline{\sigma}^a_{xx}$, *and* $\overline{\sigma}^a_{xx}$ *are known from the mechanical boundary conditions*

$$
\begin{aligned}
\overline{\sigma}^a_{xx} &= P_{xx} \\
\overline{\sigma}^b_{xx} &= 0 \\
\overline{\sigma}^b_{xy} &= 0
\end{aligned}
\tag{10.20}
$$

Table 10.1 presents the results for the case of a slender beam. The theories all yield the same values for transverse displacement and normal stress $\overline{\sigma}_{xx}$. *A seventh-order approximation is required to satisfy the mechanical boundary conditions within the accuracy of five significant digits. Results for* l/a *equal to ten and five are reported in Table 10.2. For* $l/a = 10$, *results are validated toward an elasticity solution by Lü* et al. *(2008) computed via the method presented by Sankar (2001). A fourth-order approximation is required to match the reference transverse displacement. In the case of normal and shear stress components, N should equal six and seven, respectively. For the sake of brevity, the values of the transverse shear stress and the component* $\overline{\sigma}_{xx}$ *at* $z = 0$ *and a are not reported.*

Table 10.1 Transverse displacement and stresses.

	$\overline{u}_x$ ($10^4\times$ m)	$\overline{\sigma}_{yy}$ ($10^2\times$ MPa)	$\overline{\sigma}^a_{xy}$ ($10\times$ MPa)	$\overline{\sigma}^b_{xy}$ $10\times$ (MPa)	$\overline{\sigma}^a_{xx}$ (MPa)	$\overline{\sigma}^b_{xx}$ (MPa)
$N=7$	−4.0318	26.928	−4.4586	0.0000	−1.0000	0.0000
$N=6$	−4.0318	26.928	−4.4583	−0.0008	−0.9986	0.0005
$N=5$	−4.0318	26.928	−4.4607	−0.0076	−1.0108	0.0036
$N=3$	−4.0318	26.928	−4.3508	−0.3718	−1.2152	0.0658
$N=1$	−4.0318	26.928	−2.5752	−0.8143	−0.8397	−0.0840
TBT	−4.0318	26.928	−2.5753	−0.8144	—	—
EBBT	−4.0311	26.927	—	—	—	—

Transverse displacement and stresses for $L/a = 100$.

Results are the same as in the case of a slender beam. Since the material is not homogeneous, the axial stress $\overline{\sigma}_{yy}$ no longer presents the classical Navier-type linear variation along the cross-section. The variation of $\overline{\sigma}_{yy}$ versus x at cross-section $y = L/2$ is presented in Figure 10.5. In the case of classical isotropic materials and according to Jourawsky's formula, the shear stress $\overline{\sigma}_{xy}$ varies along the cross-section quadratically, having its maximum at the cross-section center and being null for $x = 0$ and a. For the considered FGM-based structures, $\overline{\sigma}_{yy}$ deviates from that behavior as shown in Figure 10.6.

Table 10.2 Transverse displacement and stresses.

	$l/a = 10$			$l/a = 5$		
	$\overline{u}_x$ ($10^{-1}\times$ m)	$\overline{\sigma}_{yy}$ (MPa)	$\overline{\sigma}^a_{xy}$ (MPa)	$\overline{u}_x$ ($10^{-2}\times$ m)	$\overline{\sigma}_{yy}$ ($10^{-1}\times$ MPa)	$\overline{\sigma}^a_{xy}$ (MPa)
Lü *et al.* (2008)	−4.0942	26.922	−4.4527	—	—	—
$N=7$	−4.0942	26.922	−4.4527	−2.6742	67.235	−2.2171
$N=6$	−4.0942	26.922	−4.4523	−2.6742	67.237	−2.2169
$N=5$	−4.0942	26.923	−4.4548	−2.6742	67.247	−2.2183
$N=3$	−4.0938	26.942	−4.3468	−2.6731	67.434	−2.1671
$N=1$	−4.0954	26.953	−2.5694	−2.6758	67.575	−1.2760
TBT	−4.0959	26.928	−2.5753	−2.6815	67.319	−1.2876
EBBT	−4.0311	26.928	—	−2.5195	67.319	—

Transverse displacement and stresses for $l/a = 10$ and 5.

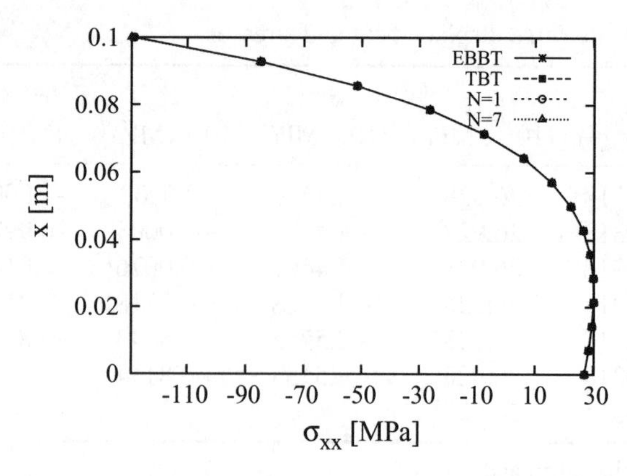

Figure 10.5 Variation of $\overline{\sigma}_{yy}$ versus x for the bidimensional problem and $y = L/2$.

Example 10.4.2 *The square cross-section as shown in Figure 10.7 is considered. Cross-section sides are equal to* 0.1 m. *A deep beams* $(L/a = 5)$ *is investigated. A sinusoidal pressure of maximal amplitude* $P_{xx} = 1$ Pa *and* $m = 1$ *is applied on the surface at* $x = a$. *The material exhibits an exponential gradation along both the x- and z-directions with* $E_0 = 1000$ MPa, α_i *such that* $E(a, 0)/E_0 = E(0, a)/E_0 = 3$, *and* $\beta_i = 0$ *with* $i = 1$ *and* 2. *The Poisson ratio is equal to* 0.3. *The higher-order CUF models are compared to a three-dimensional FEM solution (addressed as FEM 3D) developed via the commercial code MSC.Nastran. The eight-node brick element "HEXA8" is used, see Getting Started with MSC.Nastran. User's Guide (2001). The sides of the elements are* 2×10^{-3} m. *Each element is considered as homogeneous by referring to the material properties at its center point. The deformation of the cross-section at* $y = L/2$

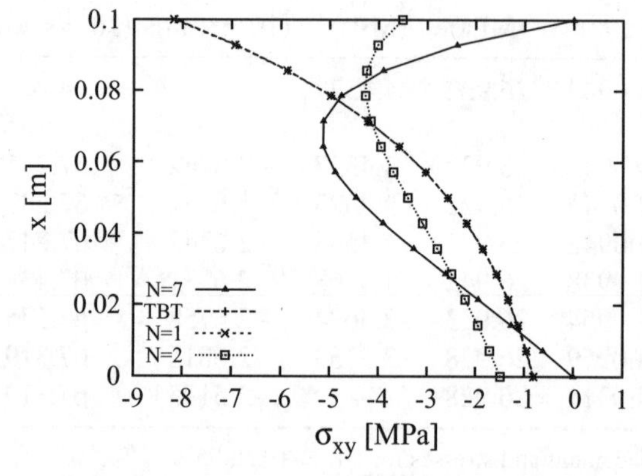

Figure 10.6 Variation of $\overline{\sigma}_{xy}$ versus x for the bidimensional problem and $y = 0$.

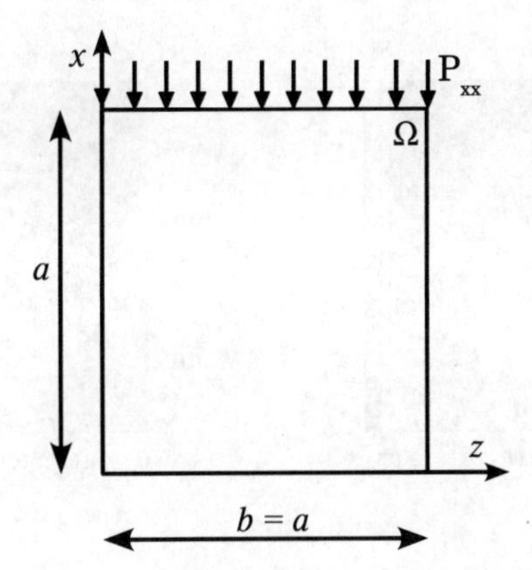

Figure 10.7 Cross-section geometry, reference system, and loading for the 3D problem.

is presented in Figure 10.8. The fourth-order model matches the reference FEM solution. The cross-section deformation is due to the Young's modulus variation law. The region of the cross-section $\{(x, z) : x/a + z/a \geq 1\}$ *is stiffer than the region* $\{(x, z) : x/a + z/a \leq 1\}$*. A first-order model yields an accurate description of the bending stress component* σ_{yy} *as shown in Figure 10.9. Results are computed at the beam's mid-span. The normal stress component* σ_{xx} *computed via the FEM 3D solution and fifth-order model is presented in Figure 10.10. A good approximation of the normal stress component* σ_{zz} *is obtained via a sixth-order model as*

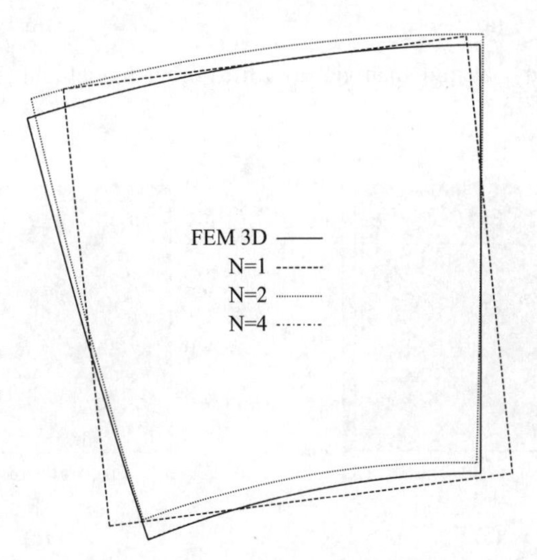

Figure 10.8 Cross-section in-plane deformation.

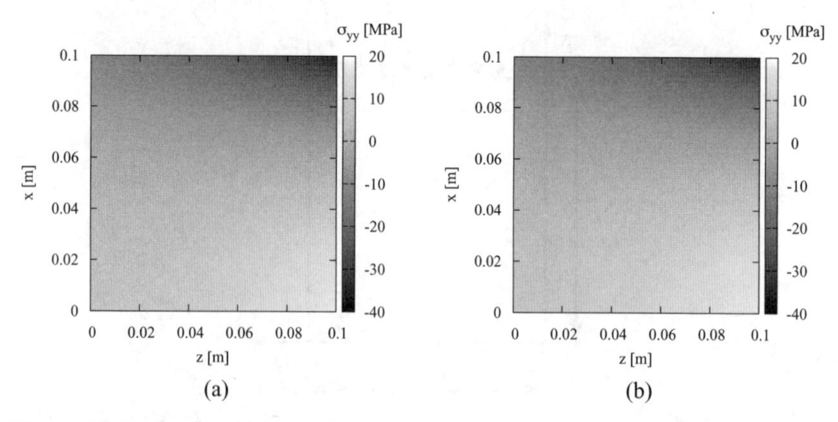

Figure 10.9 σ_{xx} at mid-span via (a) a first-order model and (b) the FEM 3D solution.

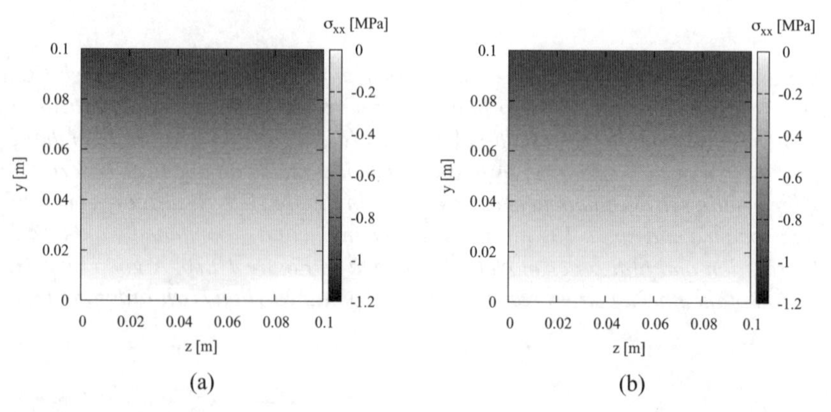

Figure 10.10 σ_{xx} at mid-span via (a) a fifth-order model and (b) the FEM 3D solution.

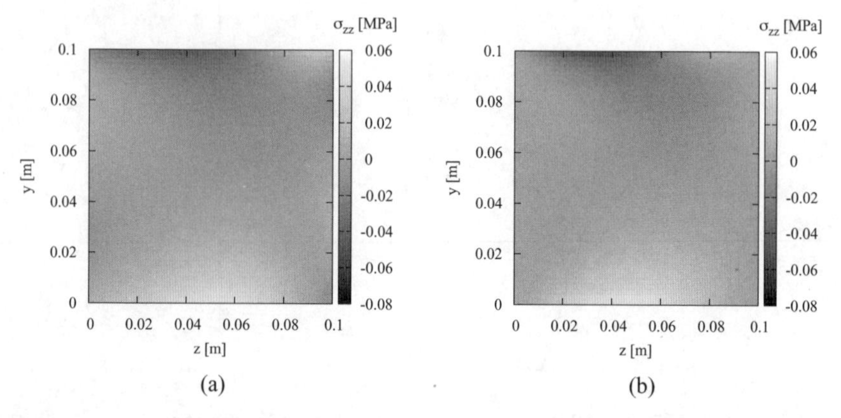

Figure 10.11 σ_{zz} at mid-span via (a) a sixth-order model and (b) the FEM 3D solution.

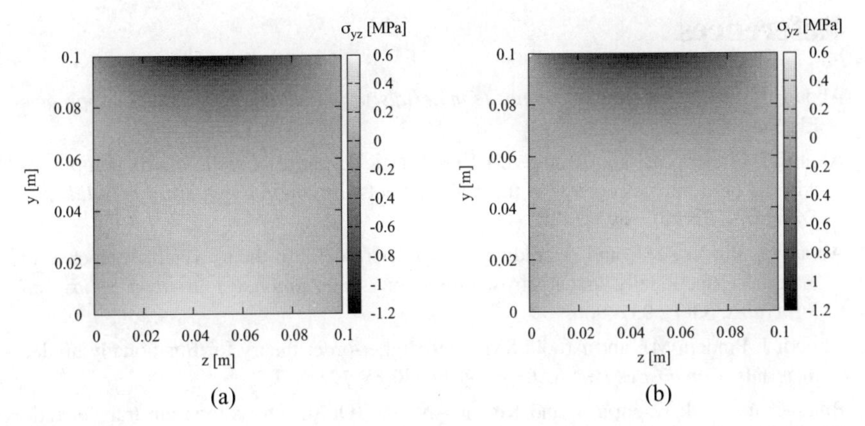

Figure 10.12 σ_{yz} at $x = 0$ via (a) a fifth-order model and (b) the FEM 3D solution.

shown in Figure 10.11. The shear stress component σ_{yz} via the FEM 3D model and a fifth-order theory is presented in Figure 10.12. Results are evaluated for $x = 0$. In Figure 10.13, the shear stress component σ_{xy} is presented. A fifth-order model has been used and compared to the reference FEM solution. As shown via a comparison with the reference FEM 3D solution, the three-dimensional stress state can be obtained by an appropriate choice of the expansion order. As far as the computational time is concerned, the proposed analytical models require less than a second, regardless of the approximation order. The FEM solution based on the proposed models, not reported here, is obtained in a few seconds for a very fine mesh. For the reference FEM 3D solution, about five minutes is required to obtain a solution.

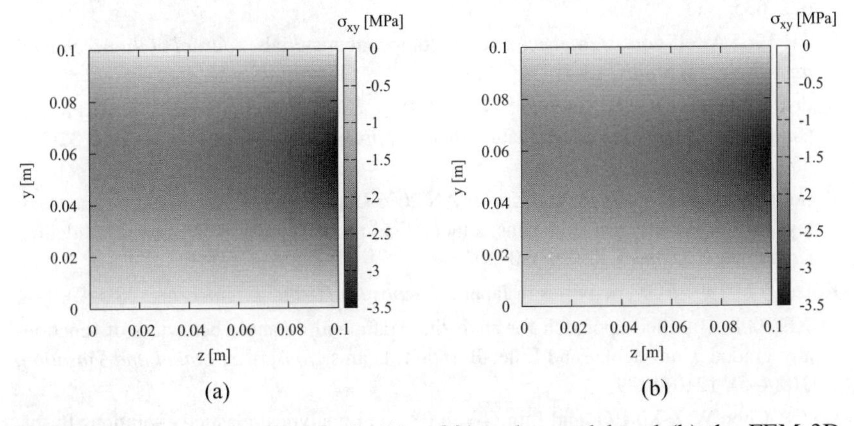

Figure 10.13 σ_{xy} at $x/l = 0$ via (a) a fifth-order model and (b) the FEM 3D solution.

References

Aboudi J 1991 *Mechanics of composite materials: a unified micromechanical approach.* Elsevier.

Aboudi J, Pindera MJ, and Arnold SM 1994 Elastic response of metal matrix composites with tailored microstructures to thermal gradients. *International Journal of Solids and Structures*, **31**(10), 1393–1428.

Aboudi J, Pindera MJ, and Arnold SM 1996 Thermoelastic theory for the response of materials functionally graded in two directions. *International Journal of Solids and Structures*, **33**(7), 931–966.

Aboudi J, Pindera MJ, and Arnold SM 1999 Higher-order theory for functionally graded materials. *Composites Part B: Engineering*, **30**(8), 777–832.

Bhattacharyya M, Kapuria S, and Kumar AN 2007 On the stress to strain transfer ratio and elastic deflection behavior for Al/SiC functionally graded material. *Mechanics of Advanced Materials and Structures*, **14**(4), 295–302.

Birman V and Byrd LW 2007 Modelling and analysis of functionally graded materials and structures. *Applied Mechanics Reviews*, **60**(5), 195–216.

Chakraborty A, Gopalakrishnan S, and Reddy JN 2003 A new beam finite element for the analysis of functionally graded materials. *International Journal of Mechanical Sciences*, **45**(3), 519–539.

Ding JH, Huang DJ, and Chen WQ 2007 Elasticity solutions for plane anisotropic functionally graded beams. *International Journal of Solids and Structures*, **44**(1), 176–196.

Finot M, Suresh S, Bull C, and Sampath S 1996 Curvature changes during thermal cycling of a compositionally graded Ni–Al_2O_3 multi-layered material. *Materials Science and Engineering A*, **205**(1–2), 59–71.

Getting Started with MSC.Nastran. User's Guide. MSC.Software Corporation, Santa Ana, CA.

Giunta G, Belouettar S, and Carrera E 2010 Analysis of FGM beams by means of classical and advanced theories. *Mechanics of Advanced Materials and Structures*, **17**(8), 622–635.

Hill R 1965 A self-consistent mechanics of composite materials. *Journal of the Mechanics and Physics of Solids*, **13**(4), 213–222.

Kadoli R, Akhtara K, and Ganesanb N 2008 Static analysis of functionally graded beams using higher order shear deformation theory. *Applied Mathematical Modelling*, **32**(12), 2509–2525.

Kapuria S, Bhattacharyya M, and Kumar AN 2008 Bending and free vibration response of layered functionally graded beams: a theoretical model and its experimental validation. *Composite Structures*, **82**(3), 390–402.

Koizumi M 1997 FGM activities in Japan. *Composites Part B: Engineering*, **28**(1–2), 1–4.

Li XF 2008 A unified approach for analyzing static and dynamic behaviors of functionally graded Timoshenko and Euler-Bernoulli beams. *Journal of Sound and Vibration*, **318**(4–5), 1210–1229.

Lü CF, Chen WQ, Xu RQ, and Lim CW 2008 Semi-analytical elasticity solutions for bi-directional functionally graded beams. *International Journal of Solids and Structures*, **45**(1), 258–275.

Miyamoto Y, Kaysser WA, Rabin BH, Kawasaki A, and Ford RG 1999 *Functionally Graded Materials: Design, Processing and Applications*. Kluwer Academic.

Mori T and Tanaka K 1973 Average stress in matrix and average elastic energy of materials with misfitting inclusions. *Acta Metallurgica*, **21**(5), 571–574.

Nemat-Nasser S and Hori M 1993 *Micromechanics: overall properties of heterogeneous materials*. North-Holland.

Sankar BV 2001 An elasticity solution for functionally graded beams . *Composites Science and Technology*, **61**(5), 689–696.

Silverman K 1964 Orthotropic beams under polynomial loads. *In: Proceedings of the ASCE, Journal of the Engineering Mechanics Division*, **90**, 293–319.

Suresh S and Mortensen A 1998 *Fundamentals of Functional Graded Materials*. IOM Communications Limited, London.

Tomota Y, Kuroki K, Mori T, and Tamura I 1976 Tensile deformation of two-ductile-phase alloys: flow curves of α-γ Fe-Cr-Ni alloys. *Materials Science and Engineering*, **24**(1), 85–94.

Wakashima HT and Tsukamoto H 1990 Micromechanical approach to the thermomechanics of ceramic-metal gradient materials. *In: 1st Symposium on Functionally Gradient Materials, The Functionally Gradient Materials Forum, Sendai, Japan*, 19–26.

Watanabe R, Nishida T, and Hirai T 2003 Present status of research on design and processing of functionally graded materials. *Metals and Materials International*, **9**(6), 513–519.

Zhu H and Sankar BV 2004 A combined Fourier Series–Galerkin method for the analysis of functionally graded beams. *Journal of Applied Mechanics*, **71**(3) 421–424.

Zuiker JR and Dvorak GJ 1994 The effective properties of functionally graded composites - I. Extension of the Mori-Tanaka method to linearly varying fields. *Composites Engineering*, **4**(1) 19–35.

11

Multi-model beam theories via the Arlequin method

In order to achieve an effective design, the mechanics of beam structures should be modeled as accurately as possible, especially in the case of non-trivial cross-sections and/or advanced materials. On the other hand, high computational costs can make refined beam theories or three-dimensional analyses unprofitable. A way to mediate between these two opposite drivers (accuracy and computational costs) is represented by a multi-model approach: "refined" and "coarse" models are duly coupled in order to describe the response of a structure with the desired accuracy and with a reasonable computational cost. After a brief discussion on some of the multi-model approaches available in the literature, the CUF multi-model approach based on the Arlequin method is presented. It is shown how to obtain a "variable kinematic" solution in which:

- high-order elements are used locally in the regions of the beam where the stress field is quasi-three-dimensional or high gradients are present; and
- low-order, computationally cheap elements are used in the remaining regions where "classical beam mechanics" is present.

Beam Structures: Classical and Advanced Theories, First Edition. Erasmo Carrera, Gaetano Giunta and Marco Petrolo.

11.1 Multi-model approaches

Many multi-model approaches have been formulated over the last few years. All these models aim at obtaining a solution that:

- reduces the computational cost (when compared to the corresponding full "refined" solution);
- is accurate; and
- does not violate the congruency and the equilibrium at the interface of the structure's sub-domains.

Many of the multi-model approaches present in the literature can be grouped into two categories: mono-theory approaches and multi-theory approaches. A brief description of some methods of each group is addressed in the following. The interested reader should refer also to the book by Reddy (2004), where the multi-model approaches are divided into multi-step or sequential methods.

11.1.1 Mono-theory approaches

As far as mono-theory approaches are concerned, the structure's sub-domains differ in the level of disretization. In the sequential adaptation methods, for instance, the structure's sub-domains differ in mesh size (*h*-adaptation, see Bank 1983) or degrees of freedom of the shape functions (*p*-adaptation, see Szabo and Babuska 1992) or both (*hp*-adaptation, see Bathe 1996). Mesh size and shape functions are modified according to a sequential approach based on the iteration of analysis and error estimation. In the multi-grid method (see Fish *et al.* 1996), coarse and fine meshes share information inside an iterative algorithm. In the extended finite element method by Möes *et al.* (1999), the basis of the shape functions is enriched to account for the discontinuity of the displacement field.

11.1.2 Multi-theory approaches

In the case of multi-model methods, the structure's sub-domains differ in the kinematic assumptions. The very theoretical models describing the mechanics of a structure are adapted. In the *s*-version method, see Fish (1992) and Fish and Markolefas (1993), incompatible meshes (different element size and polynomial order) with a local–global border are coupled. Park and Felippa (2000) presented a continuum-based variational principle for the formulation of discrete governing equations of partitioned structural systems, including coupled substructures as well as sub-domains obtained by mesh decomposition. A variational approach to couple kinematically incompatible structural models was presented by Blanco *et al.* (2008). In the three-field formulation by Brezzi and Marini (2005), an additional grid is introduced at the interface. The unknowns are represented independently

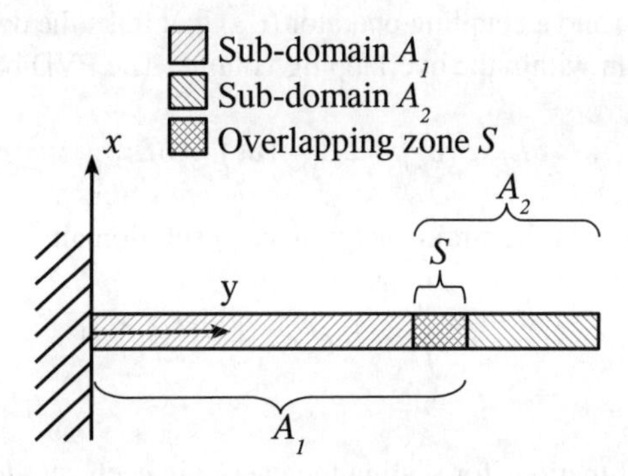

Figure 11.1 Beam structure divided into two overlapping sub-domains.

in each sub-domain and on the interface, the matching being provided by suitable Lagrange multipliers. The Arlequin method was proposed by Ben Dhia (1998, 1999 and Ben Dhia and Rateau (2005). The coupling among different numerical models is obtained through Lagrange multipliers. This method was adopted by Hu *et al.* (2008, 2010) for the linear and non linear analysis of sandwich beams modeled via one- and two-dimensional finite elements and, within the CUF, by Biscani *et al.* (2011) for the analysis of isotropic and composite beams.

11.2 The Arlequin method in the context of the unified formulation

The beam volume V is considered as the union of two partially overlapping sub-domains A_1 and A_2 as shown in Figure 11.1:

$$\begin{aligned} V &= A_1 \cup A_2 \\ S &= A_1 \cap A_2 \end{aligned} \tag{11.1}$$

S being the overlapping volume. For each sub-domain, a different generalized expansion of the displacement fields is assumed:

$$\mathbf{u}_\xi = N_i F_{\tau_\xi} \mathbf{q}_{\tau_\xi i} \quad \text{with } \tau_\xi = 1, 2, \ldots, N_u^{A_\xi}, \quad \xi = 1, 2 \tag{11.2}$$

where ξ is a dummy index that counts the sub-domains. The global mechanical problem is solved by merging together the two sub-domains via the Arlequin method. The internal and external virtual works are computed for each sub-domain. The structural integrity in the overlapping volume is ensured via a Lagrangian multiplier field ($\boldsymbol{\lambda}$) and a coupling operator (C_ξ) that links the degrees of freedom of each sub-domain within the overlapping volume. The PVD becomes

$$\delta L_{in,\xi}\left(\mathbf{u}_\xi\right) + \delta L_{c,\xi}\left(\mathbf{u}_\xi\right) = \delta L_{ex,\xi} \tag{11.3}$$

The virtual variation of the strain energy in each sub-domain is

$$\delta L_{in,\xi} = \int_{A_\xi} \alpha_\xi \left(\delta\boldsymbol{\epsilon}_n^T \boldsymbol{\sigma}_n + \delta\boldsymbol{\epsilon}_p^T \boldsymbol{\sigma}_p\right) dV \tag{11.4}$$

α_ξ are weighting functions for scaling the energy in each sub-domain in order to not consider the energy in the overlapping volume twice:

$$\begin{aligned} &\alpha_\xi = 1 \;\; \forall (x, y, z) \in A_\xi \setminus S \\ &\alpha_1 + \alpha_2 = 1 \;\; \forall (x, y, z) \in S \end{aligned} \tag{11.5}$$

According to Ben Dhia (1999), the weighting functions should be such that the sub-domain with a more accurate description has a higher weight in the global equilibrium. Unless stated otherwise, a constant value equal to 0.99 is assumed for the sub-domain in which the refined model is adopted. The virtual external work is treated in a similar manner. $\delta L_{c,\xi}$ is the virtual coupling work:

$$\delta L_{c,\xi} = (-1)^\xi \, \delta C_\xi \left(\boldsymbol{\lambda}, \mathbf{u}_\xi\right) \tag{11.6}$$

where C_ξ is a coupling operator. Two coupling operators are considered, see Hu *et al.* (2008):

- L^2 coupling:

$$\delta C_\xi = \int_{S_\xi} \delta\boldsymbol{\lambda}^T \mathbf{u}_\xi \, dV \tag{11.7}$$

- H^1 coupling:

$$\delta C_\xi = \int_{S_\xi} \left\{\delta\boldsymbol{\lambda}^T \mathbf{u}_\xi + \tilde{l}^2 \left[\delta\boldsymbol{\epsilon}_n^T (\boldsymbol{\lambda}) \boldsymbol{\epsilon}_n \left(\mathbf{u}_\xi\right) + \delta\boldsymbol{\epsilon}_p^T (\boldsymbol{\lambda}) \boldsymbol{\epsilon}_p \left(\mathbf{u}_\xi\right)\right]\right\} dV \tag{11.8}$$

$\tilde{l}$ is a real parameter representative of a characteristic length. $\boldsymbol{\epsilon}(\boldsymbol{\lambda})$ is defined in the same manner as the mechanical strain $\boldsymbol{\epsilon}\left(\mathbf{u}_\xi\right)$ where the Lagrangian multiplier field is used instead of the displacement one. The Lagrangian multiplier is discretized according to the CUF generalized expansion:

$$\boldsymbol{\lambda} = N_i F_{\tau_\lambda} \boldsymbol{\Lambda}_{\tau_\lambda i} \tag{11.9}$$

where $\mathbf{\Lambda}_{\tau_\lambda i}$ is the nodal unknown vector. The fundamental nucleus of the coupling matrix $\mathbf{C}_\xi^{ij\tau_\xi s_\lambda}$ is derived coherently to the weak form of the governing equations

$$\delta L_{in} = \delta \mathbf{q}_{sj}^T \mathbf{K}^{ij\tau s} \mathbf{q}_{\tau i} \tag{11.10}$$

by substituting of Equation 11.9 into Equation 11.7 or 11.8:

$$\delta C_\xi = \delta \mathbf{\Lambda}_{s_\lambda j}^T \mathbf{C}_\xi^{ij\tau_\xi s_\lambda} \mathbf{q}_{\tau_\xi i} \tag{11.11}$$

In the case of L^2 coupling, the fundamental nucleus is diagonal and its components are

$$C_{\xi mn}^{ij\tau_\xi s_\lambda} = \delta_{nm} I_{ij} J_{\tau_\xi s_\lambda} \quad \text{with } m, n = x, y, z \tag{11.12}$$

where δ_{nm} is Kronecker's delta. The terms I_{ij} are the integrals of the shape function along the axis of an element

$$I_{ij} = \int_l N_i N_j \, dy \tag{11.13}$$

and the terms $J_{\tau_\xi s_\lambda}$ are the cross-section moments

$$J_{\tau_\xi s_\lambda} = \int_\Omega F_{\tau_\xi} F_{s_\lambda} \, d\Omega \tag{11.14}$$

As far as the coupling operator H^1 is concerned, the fundamental nucleus of the coupling matrix can be obtained straightforwardly by noticing that the H^1 coupling operator is the sum of the L^2 one and a term similar to the virtual internal work:

$$\delta L^{int} = \int_l \int_\Omega \left(\delta \boldsymbol{\epsilon}_n^T \boldsymbol{\sigma}_n + \delta \boldsymbol{\epsilon}_p^T \boldsymbol{\sigma}_p \right) d\Omega \, dx \tag{11.15}$$

The components of this latter term are those of the stiffness matrix that correspond to the diagonal terms of the constitutive matrices $\tilde{\mathbf{C}}_{pp}$ and $\tilde{\mathbf{C}}_{nn}$:

$$\begin{aligned}
C_{\xi xx}^{ij\tau_\xi s_\lambda} = C_{\xi yy}^{ij\tau_\xi s_\lambda} = C_{\xi zz}^{ij\tau_\xi s_\lambda} &= I_{ij} J_{\tau_\xi s_\lambda} + \bar{l}^2 \left[I_{i,y j,y} J_{\tau_\xi s_\lambda} + I_{ij} \left(J_{\tau_{\xi,x} s_{\lambda,x}} + J_{\tau_{\xi,z} s_{\lambda,z}} \right) \right] \\
C_{\xi xy}^{ij\tau_\xi s_\lambda} = \bar{l}^2 I_{i,y j} J_{\tau_\xi s_{\lambda,x}} \qquad & C_{\xi yx}^{ij\tau_\xi s_\lambda} = \bar{l}^2 I_{ij,y} J_{\tau_{\xi,x} s_\lambda} \\
C_{\xi yz}^{ij\tau_\xi s_\lambda} = \bar{l}^2 I_{i,y j} J_{\tau \xi s_{\lambda,z}} \qquad & C_{\xi zy}^{ij\tau_\xi s_\lambda} = \bar{l}^2 I_{ij,y} J_{\tau_{\xi,z} s_\lambda} \\
C_{\xi xz}^{ij\tau_\xi s_\lambda} = \bar{l}^2 I_{ij} J_{\tau_{\xi,x} s_{\lambda,z}} \qquad & C_{\xi zx}^{ij\tau_\xi s_\lambda} = \bar{l}^2 I_{ij} J_{\tau_{\xi,z} s_{\lambda,x}}
\end{aligned} \tag{11.16}$$

where

$$I_{i_{(,y)}j_{(,y)}} = \int_l N_{i_{(,y)}} N_{j_{(,y)}} \, dy \tag{11.17}$$

and

$$J_{\tau_{\xi(,x)(,z)} s_{\lambda(,x)(,z)}} = \int_\Omega F_{\tau_{\xi(,x)(,z)}} F_{s_{\lambda(,x)(,z)}} \, d\Omega \tag{11.18}$$

As suggested by Ben Dhia (1999), the same approximation order should be assumed for the low-order model and the Lagrangian multiplier. Considering the whole structure and assuming that the refined model is adopted in the sub-domain A_2 (see Figure 11.1)

$$N_u^{A_2} \geq N_u^{A_1} \tag{11.19}$$

the governing equations of the variable kinematic problem for the whole beams are

$$\begin{bmatrix} \overline{\mathbf{K}}_{A_1\setminus S}^{ij\tau_1 s_1} & \mathbf{0} & \mathbf{0} & \mathbf{0} & \mathbf{0} \\ \mathbf{0} & (1-\alpha)\,\overline{\mathbf{K}}_{A_1\cap S}^{ij\tau_1 s_1} & \mathbf{0} & \mathbf{0} & \overline{\mathbf{C}}_1^{ij\tau_1 s_1 T} \\ \mathbf{0} & \mathbf{0} & \overline{\mathbf{K}}_{A_2\setminus S}^{ij\tau_2 s_2} & \mathbf{0} & \mathbf{0} \\ \mathbf{0} & \mathbf{0} & \mathbf{0} & \alpha\overline{\mathbf{K}}_{A_2\cap S}^{ij\tau_2 s_2} & -\overline{\mathbf{C}}_2^{ij\tau_2 s_1 T} \\ \mathbf{0} & \overline{\mathbf{C}}_1^{ij\tau_1 s_1} & \mathbf{0} & -\overline{\mathbf{C}}_2^{ij\tau_2 s_1} & \mathbf{0} \end{bmatrix} \begin{Bmatrix} \overline{\mathbf{q}}_{\tau_1 i}^{A_1\setminus S} \\ \overline{\mathbf{q}}_{\tau_1 i}^{A_1\cap S} \\ \overline{\mathbf{q}}_{\tau_2 i}^{A_2\setminus S} \\ \overline{\mathbf{q}}_{\tau_2 i}^{A_2\cap S} \\ \overline{\mathbf{\Lambda}}_{\tau_1 i} \end{Bmatrix} = \begin{Bmatrix} \overline{\mathbf{P}}_{s_1 j}^{A_1\setminus S} \\ (1-\alpha)\,\overline{\mathbf{P}}_{s_1 j}^{A_1\cap S} \\ \overline{\mathbf{P}}_{s_2 j}^{A_2\setminus S} \\ \alpha\overline{\mathbf{P}}_{s_2 j}^{A_2\cap S} \\ 0 \end{Bmatrix} \tag{11.20}$$

where $\overline{\mathbf{K}}_{A_\xi\setminus S}^{ij\tau_\xi s_\xi}$ are the global stiffness matrices for the two uncoupled sub-domains and $\overline{\mathbf{K}}_{A_\xi\cap S}^{ij\tau_\xi s_\xi}$ are the global stiffness matrices for the overlapping volumes, and $\overline{\mathbf{q}}_{\tau_\xi i}^{A_\xi\setminus S}$ and $\overline{\mathbf{q}}_{\tau_\xi i}^{A_\xi\cap S}$ are the uncoupled and coupled generalized nodal unknowns, respectively. $\overline{\mathbf{P}}_{s_\xi j}^{A_\xi\setminus S}$ and $\overline{\mathbf{P}}_{s_\xi j}^{A_\xi\cap S}$ are variationally consistent, generalized, external nodal loads.

11.3 Examples

Two numerical examples are presented to show the accuracy of the multi-model approach. A square and a thin-walled I-shaped cross-section are considered. In both the examples, the load is localized at the beam's mid-span, the beams are simply supported, and the same isotropic material is used. Multi-model solutions are compared to the corresponding mono-model strong and weak solutions and 3D FEM solutions obtained via the commercial code Abaqus.

Example 11.3.1 *A simply supported beam with a square cross-section is considered, see Figure 11.2. The sides length is* 0.2 m *and the slenderness ratio* L/a *is equal to* 10. *The figure presents also the points on the cross-section where displacements and stresses are evaluated. The beam is made of an isotropic material whose properties are: Young's modulus* 75 GPa *and Poisson ratio* 0.3. *The beam undergoes a localized uniform pressure* P_{xx} *equal to* 1 Pa *acting on* 10% *of the length and centered at mid-span. The loading is applied on the top of the cross-section as shown in Figure 11.3. Results are given in terms of the following dimensionless displacements:*

$$\begin{aligned}\overline{u}_x &= \frac{4Ea}{L^2P_{xx}}\,u_x\left(\frac{a}{2},0,0\right)\\ \overline{u}_y &= \frac{4Ea}{L^2P_{xx}}\,u_y\left(-\frac{a}{2},\frac{L}{2},\frac{a}{2}\right)\\ \overline{u}_z &= \frac{4Ea}{L^2P_{xx}}\,u_z\left(-\frac{a}{2},\frac{L}{2},\frac{a}{2}\right)\end{aligned} \tag{11.21}$$

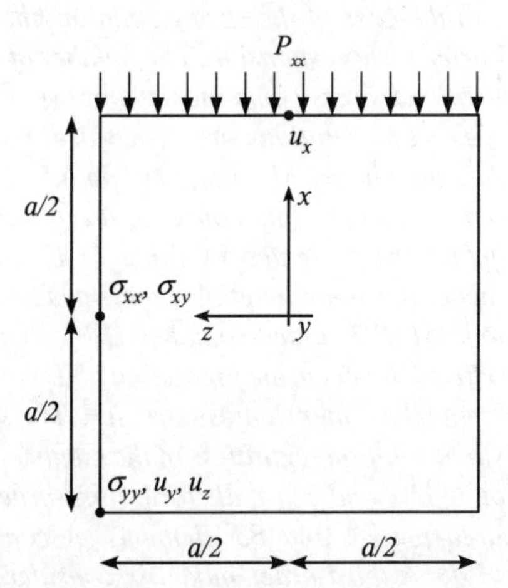

Figure 11.2 Square cross-section geometry, loading, and points where the displacements and stresses are evaluated.

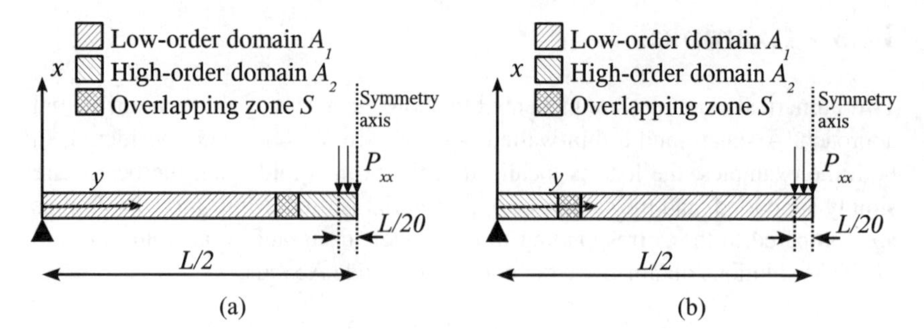

Figure 11.3 Simply supported beams under a localized uniform pressure P_{xx}. Arlequin model with refined elements near (a) the loading application zone or (b) the constraint.

and stresses, see also Figure 11.2:

$$
\begin{aligned}
\overline{\sigma}_{yy} &= \frac{1}{P_{xx}}\,\sigma_{yy}\left(-\frac{a}{2},\frac{L}{2},\frac{a}{2}\right)\\
\overline{\sigma}_{xy} &= \frac{1}{P_{xx}}\,\sigma_{xy}\left(0,0,\frac{a}{2}\right)\\
\overline{\sigma}_{xx} &= \frac{1}{P_{xx}}\,\sigma_{xx}\left(0,\frac{L}{2},\frac{a}{2}\right)
\end{aligned}
\tag{11.22}
$$

Only half of the structure is investigated due to the symmetry of the problem, see Figure 11.3. Each weak form solution is first validated toward the corresponding strong form, Navier-type solution, and a 3D FEM solution obtained via the commercial code Abaqus. In the case of the strong solution, the localized loading is approximated via a Fourier series expansion. The number of approximation terms is such that displacement and stress components converge up to three significant figures. As far as the 3D FEM solutions are concerned, the quadratic C3DR20 element is used, see Abaqus Theory Manual, version 6.5. The mesh is such that the maximum displacement components converge up to four significant figures. Displacement components are presented in Table 11.1. Results are computed considering 20 *elements of the same length* l. *Elements can have two (B2), three (B3), and four (B4) nodes. EBBT underestimates* $\overline{u}_x$ *by about* 1.2% *since it does not account for shear effects. An accurate prediction of* $\overline{u}_z$ *calls for a second-order theory. Table 11.2 presents the dimensionless stresses. The stress component* $\overline{\sigma}_{yy}$ *converges to the analytical solution regardless of the number of nodes per element. An accurate evaluation of* $\overline{\sigma}_{xy}$ *and* $\overline{\sigma}_{xx}$ *calls for a third-order theory at least and a finer mesh is required for B2 and B3 elements. A convergence analysis is presented in Figure 11.4. A third-order model is considered. Ten elements are sufficient in the case of B4 elements. After proving the accuracy of the mono-model CUF FEM solutions, the multi-model Arlequin-based approach is used.*

Table 11.1 Dimensionless displacements.

	$10 \times \overline{u}_y$			$\overline{u}_x$			$10^2 \times \overline{u}_z$		
3D FEM	3.742			2.544			2.125		
CUF	B2	B3/4	SS*	B2	B3/4	SS	B2	B3/4	SS
$N = 3$	3.740	3.741	3.741	2.541	2.544	2.544	2.111	2.113	2.113
$N = 2$	3.734	3.736	3.736	2.531	2.533	2.533	2.099	2.100	2.106
$N = 1$	3.747	3.749	3.749	2.549	2.522	2.522	−0.055	−0.053	−0.053
TBT	3.735	3.737	3.737	2.546	2.549	2.549	0.000	0.000	0.000
EBBT	3.736	3.737	3.737	2.484	2.487	2.487	0.000	0.000	0.000

*strong solution.
Dimensionless displacements via mono-model solutions.

Elements based on a first-order theory (low-order model) are coupled to those based on a fourth-order one (refined model). Two configurations are considered. In the first one, identified as "MM-A^a," the refined sub-domain is near the loading application area. For the second configuration (MM-B^a), the refined sub-domain is near the constraint as shown in Figure 11.3. Two different configurations are addressed since the load application area and constrained regions are likely to present a three-dimensional stress field. Due to the symmetry of the problem, only

Table 11.2 Dimensionless stresses.

	$-10^{-1} \times \overline{\sigma}_{xx}$			$10^1 \times \overline{\sigma}_{xy}$			$10^1 \times \overline{\sigma}_{yy}$		
FEM 3D	1.428			8.595			5.278		
				Strong form CUF solution					
$N = 4$	1.425			8.462			5.208		
$N = 3$	1.432			8.462			5.245		
$N = 1$	1.446			5.000			4.871		
TBT	1.425			5.000			–		
EBBT	1.425			–			–		
				Weak form CUF solution					
	B2	B3	B4	B2	B3	B4	B2	B3	B4
$N = 4$	1.410	1.427	1.425	8.100	8.581	8.461	6.015	5.192	5.214
$N = 3$	1.423	1.436	1.432	8.100	8.581	8.461	5.998	5.235	5.250
$N = 1$	1.437	1.450	1.446	4.639	5.120	5.000	5.020	4.865	4.870
TBT	1.415	1.428	1.425	4.639	5.120	5.000	–	–	–
EBBT	1.415	1.428	1.425	–	–	–	–	–	–

Dimensionless stresses via mono-model solutions.

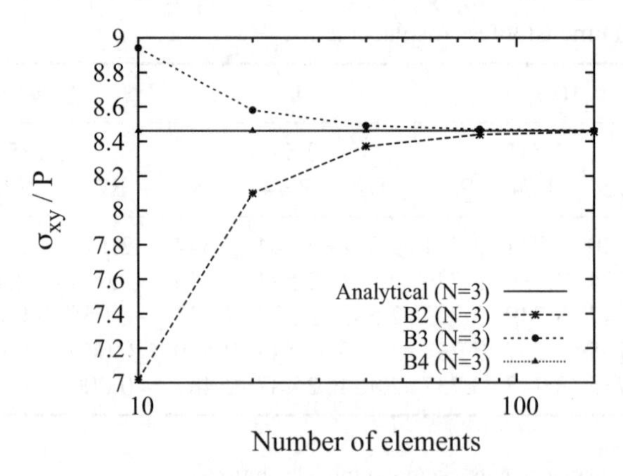

Figure 11.4 Dimensionless stress $\overline{\sigma}_{xy}$ at $x = 0$ versus the number of elements for square cross-section beam, $l/a = 10$, third-order model.

half of the structure is investigated and only a superimposition volume is needed. In the general case in which the location of a three-dimensional stress field cannot be determined a priori, refined sub-domains should be chosen on the basis of experience and preliminary analyses via low-order models. Analyses have been carried out considering both L^2 and H^1 coupling. For the latter, different values of $\tilde{l}$ in Equation 11.8 have been considered. Unless stated otherwise, the L^2 coupling is therefore used. The coarse sub-domain A_1 is meshed via sixteen first-order B4 elements whereas five fourth-order B4 elements are considered for the refined sub-domain A_2. A superimposed element is considered in the overlapping volume since, in general, the number of superimposed elements does not affect the accuracy, but only the total number of degrees of freedom (DOFs) In order to reduce the number of DOFs, the best choice consists of a superimposed element only. Displacements and stresses for a deep beam are reported in Table 11.3.

Table 11.3 Variable kinematic models.

	$10^1 \times \overline{u}_x$	$\overline{u}_y$	$10^2 \times \overline{u}_z$	$-10^{-1} \times \overline{\sigma}_{xx}$	$10^1 \times \overline{\sigma}_{xy}$	$10^1 \times \overline{\sigma}_{yy}$	DOFs*
$N = 4$	3.741	2.544	2.118	1.425	8.461	5.214	2745
$N = 1$	3.749	2.522	−0.053	1.446	5.000	4.870	549
MM-A[a]	3.729	2.537	2.116	1.424	5.000	5.217	1197
MM-A[b]	3.716	2.547	−0.056	1.444	8.352	4.807	1197

[a] Refined elements near the loading application zone.
[b] Refined elements near the simple support.
* DOFs: degrees of freedom.
Displacements and stresses computed via mono- and multi-model solutions.

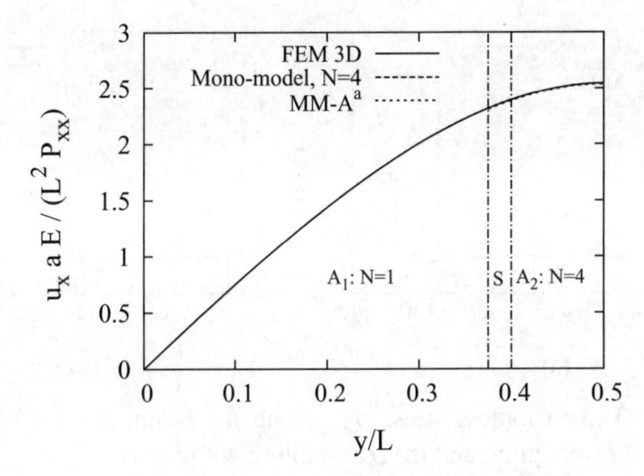

Figure 11.5 Dimensionless displacement $\overline{u}_x$ along the beam axis.

Results are aligned with those obtained through mono-theory models having the same expansion order, proving the effectiveness of the Arlequin method in coupling domains having finite elements based on theories with different expansion order. A comparison between the considered variable kinematic models and the fourth-order mono-model shows that the total number of DOFs is reduced by more than half. Figures 11.5 to 11.7 show the variation along the beam axis of $\overline{u}_x$, $\overline{u}_z$, *and* $\overline{\sigma}_{xx}$, *respectively. The MM-A*^{a} *solution is compared to a mono-model solution with* 20 *fourth-order elements identified in the figures as "Reference." In the overlapping volume S, two solutions exist. Their values do not necessarily match. Global bending response* u_x *is accurately described by both first- and fourth-order models. In the case of* u_z, *first-order theory does not account for the warping of the section, whereas fourth-order elements match the reference*

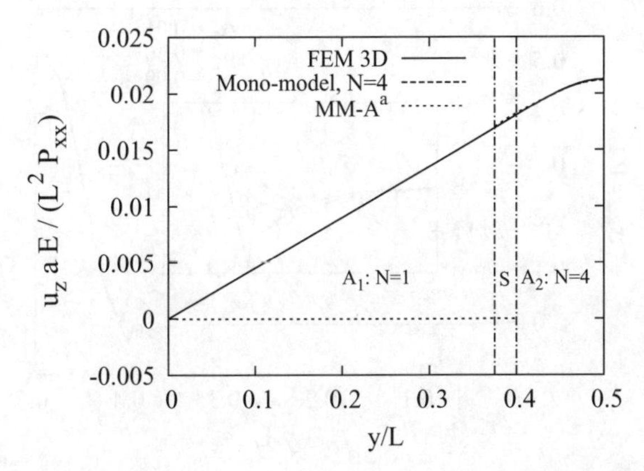

Figure 11.6 Dimensionless displacement $\overline{u}_z$ along the beam axis.

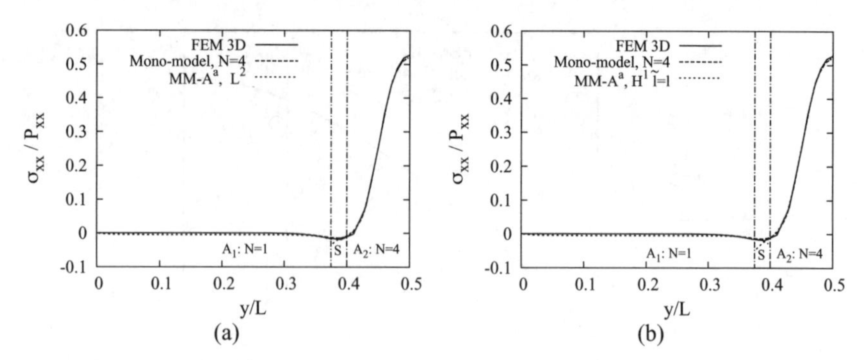

Figure 11.7 Dimensionless stress $\overline{\sigma}_{xx}$ along the beam axis for square cross-section via (a) L^2 coupling and (b) H^1 coupling with $\tilde{l} = l$.

solution. In the case of σ_{xx}, both L^2 and H^1 coupling operators have been considered. In the case of H^1 coupling, the parameter $\tilde{l}$ is equal to element length l. *Results differ slightly in the superimposed volume only. They do not change with increasing $\tilde{l}$. The H^1 solution converges to the L^2 one once $\tilde{l}$ decreases. As far as the sensitivity of the Arlequin method upon the weighting functions α_ξ in Equation 11.4 is concerned, Figure 11.8 shows the influence of α_2 on σ_{xx}. α_2 influences the solution in the coupling domain only. With increasing α_2, σ_{xx} becomes smoother and smoother since the refined model assumes more and more relevance. Outside the coupling domain, α_2 does not affect the solution. A qualitative comparison of the stress component $\overline{\sigma}_{xy}$ via the three-dimensional FEM solution, fourth-order model, and MM-A^b solution is presented in Figure 11.9. The results are in good agreement.*

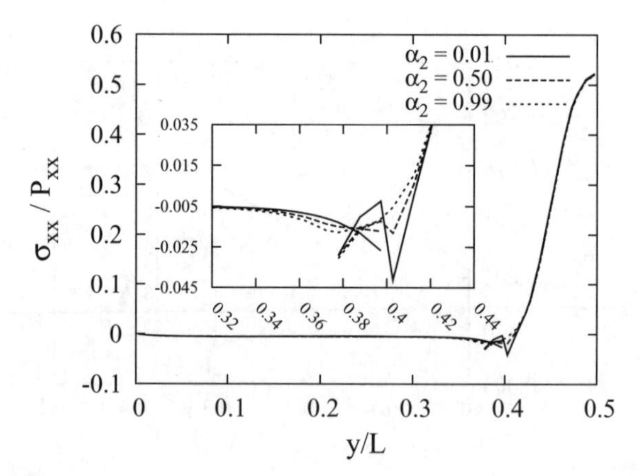

Figure 11.8 Dimensionless stress $\overline{\sigma}_{xx}$ along the beam axis with varying α_2.

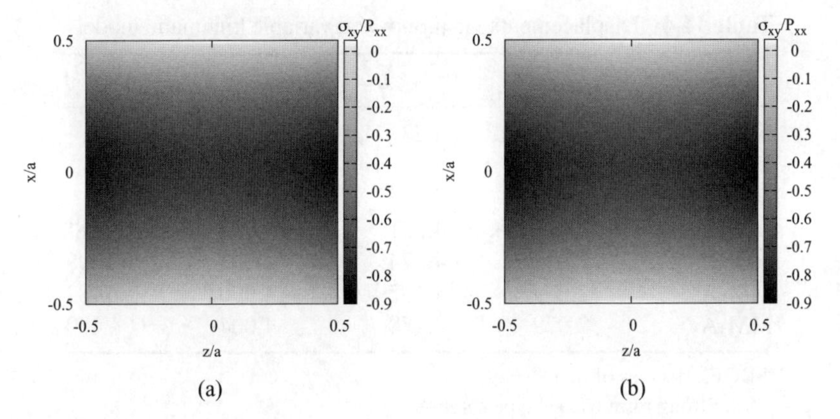

Figure 11.9 Dimensionless stress $\overline{\sigma}_{xy}$ above the cross-section at $y = 0$ via (a) a fourth-order mono-model solution and (b) the MM-A^b solution.

Example 11.3.2 *A beam with a thin-walled I-shaped cross-section is considered. The wall thickness is equal to* 10% *of a main cross-section dimension a equal to* 0.2 m. *A moderately deep beam,* $L/a = 30$, *is considered. Cross-section geometry and loading are presented in Figure 11.10. The beam is made of the same material as in Example 11.3.1. It is also loaded and constrained as in the previous example. The displacements and the stresses are put in a dimensionless form according to Equations 11.21. The points where displacements and stresses are evaluated are presented in Figure 11.10. Results are assessed versus the corresponding mono-model weak and strong form Navier-Type solutions and 3D FEM solutions via Abaqus. Four-node elements are considered. The displacements are presented*

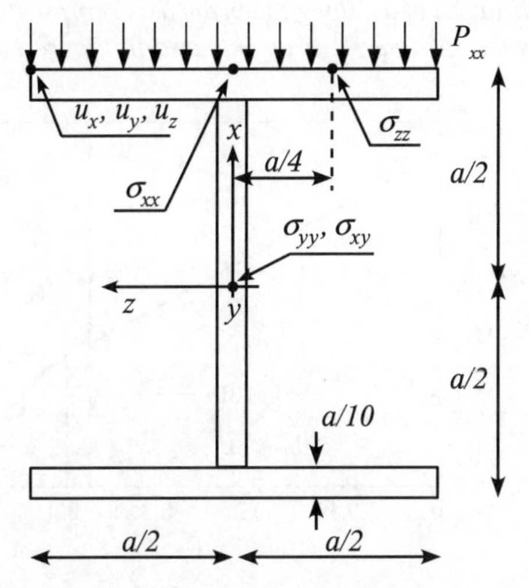

Figure 11.10 I-shaped cross-section geometry, loading, and verification points.

Table 11.4 Displacements via mono- and variable kinematic models.

	$-\overline{u}_y$	$10^{-1} \times \overline{u}_x$	$-10^2 \times \overline{u}_z$	DOFs*
FEM 3D	2.082	4.233	5.216	—
$N = 15$ SS[†]	2.078	4.231	4.993	—
$N = 2$ SS	2.078	4.174	4.003	—
$N = 15$	2.073	4.231	4.992	24 888
$N = 2$	2.073	4.174	4.003	1098
MM-A[a]	2.079	4.192	4.990	7482
MM-A[b]	2.079	4.188	4.004	7482

* DOFs: degrees of freedom.
[†] SS: Strong form Navier-type solution.
[a] Refined elements near the loading application zone.
[b] Refined elements near the simple support.
Displacements with the Arlequin method compared to mono-theory models.

in Table 11.4. The weak solution converges to the strong one. Displacement $\overline{u}_y$ is accurately modeled by classical theories. The displacement component $\overline{u}_z$ is due to the localized loading and an accurate prediction calls for high-order theories. An expansion order as high as 15 is considered. A fourth-order model (not presented here for the sake of brevity) underestimates it by about 20%, whereas the difference is about 4% in the case of a 15th-order theory. As far as the multi-model solutions are concerned, the coarse sub-domain A_1 is meshed via 16 second-order elements whereas five 15th-order elements are considered for sub-domain A_2. The Arlequin method proves to be effective in merging sub-domains having different finite elements. The total number of degrees of freedom in the analysis is reduced to less than a third. The variation of $\overline{u}_z$ along the beam axis is presented in Figure 11.11. The coupling operators L^2 and H^1 are compared in Figure 11.12. Results differ

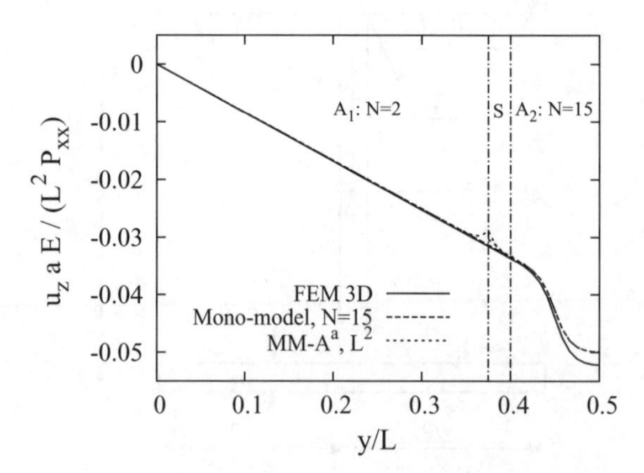

Figure 11.11 Dimensionless displacement $\overline{u}_z$ along the axis of the beam via the L^2 coupling.

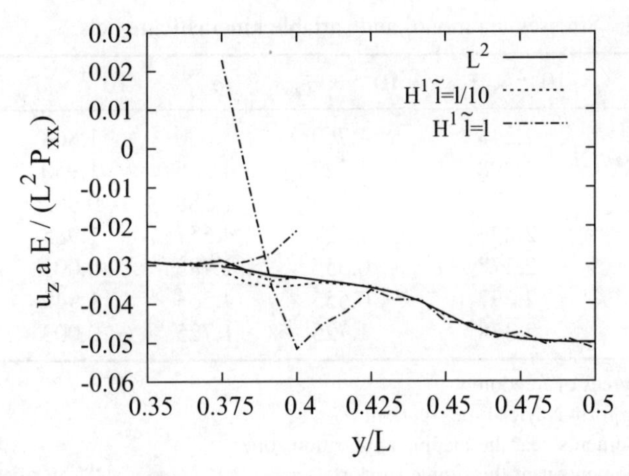

Figure 11.12 Coupling operator comparison for dimensionless displacement $\overline{u}_z$ along the axis of the beam.

significantly in the superimposed volume and in its neighborhood with increasing $\tilde{l}$. The solutions become coincident moving away from the coupling domain. The deformed section at mid-span is presented in Figure 11.13. The deformed section computed via mono-model 15th-order theory and the variable kinematic model differ mainly by a rigid translation. The Arlequin method captures local phenomena

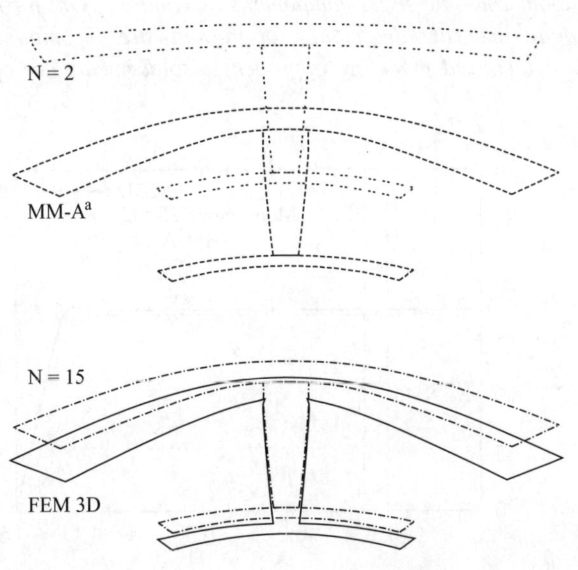

Figure 11.13 Deformed cross-section at mid-span in the case of I-shaped cross-section beam, $l/a = 30$.

Table 11.5 Stresses via mono- and variable kinematic models.

	$10^{-2} \times \overline{\sigma}_{yy}$	$10^{-1} \times \overline{\sigma}_{xy}$	$\overline{\sigma}_{xx}$	$10^{-1} \times \overline{\sigma}_{zz}$	DOFs*
FEM 3D	2.244	1.769	5.090	1.868	–
$N = 15$ SS[†]	2.298	1.727	4.565	1.950	–
$N = 2$ SS	2.378	0.555	1.788	0.002	–
$N = 15$	2.294	1.727	4.557	1.985	24 888
$N = 2$	2.378	0.555	1.785	0.003	1098
Arlequin[a]	2.297	0.555	4.563	1.949	7482
Arlequin[b]	2.378	1.725	1.725	0.003	7482

* DOFs: degrees of freedom.
[†] SS: Strong form Navier-type solution.
[a] Refined elements near the loading application zone.
[b] Refined elements near the simple support.
Stresses with the Arlequin method compared to mono-theories models.

such as the absolute value of $\overline{u}_z$ and the variation of $\overline{u}_x$ along the z-axis, which are responsible for the shape of the deformed section. If a quantity in the high-order part of the model strongly depends upon its value in the low-order part where it is not correctly modeled, the inaccuracy propagates from the low- to the high-order part. This is the case for the absolute value of $\overline{u}_x$ that is responsible for the position of the deformed section. The dimensionless stresses are reported in Tables 11.5. $\overline{\sigma}_{yy}$ and $\overline{\sigma}_{xy}$ evaluated via a 15th-order model differ from the reference 3D FEM solution by about 2%. The stress components $\overline{\sigma}_{zz}$ and $\overline{\sigma}_{xx}$ call for higher-order models. A more accurate description of normal stress components can be obtained via a localized modeling approach: displacements are approximated

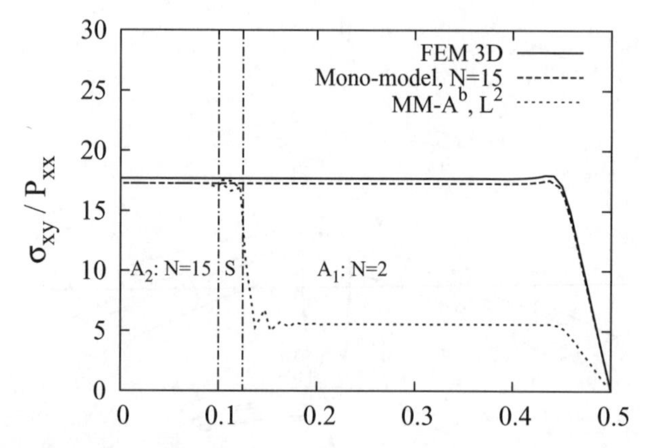

Figure 11.14 Dimensionless stress $\overline{\sigma}_{xy}$ along the axis of the beam.

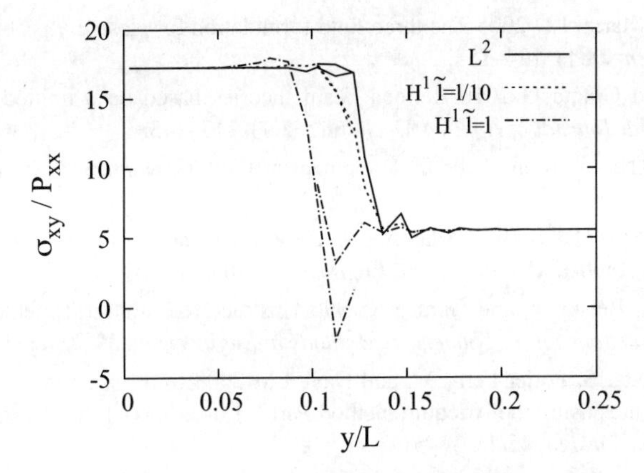

Figure 11.15 Coupling operator comparison for dimensionless stress $\overline{\sigma}_{xy}$ along the axis of the beam.

in each cross-section sub-domains by polynomial functions, such as those of Lagrange or Legendre, that ensure the congruency of the displacement fields within sub-domains' shared borders. The variable kinematic solutions yield good results. Figure 11.14 shows the variation of $\overline{\sigma}_{xy}$ along the beam axis. The oscillations in the coupling zone yield small oscillations in its neighborhood. They depend upon the coupling operator as shown in Figure 11.15. The H^1 coupling operator yields a smoother solution. Nevertheless, high values of $\tilde{l}$ cause a loss of accuracy.

References

Abaqus Theory Manual, version 6.5. Hibbit and Karlson and Sorensen Inc, Pawtucket, RI.

Bank RE 1983 The efficient implementation of local mesh refinement algorithms. *In: Adaptive Computational Methods for Partial Differential Equations*, 74–81. SIAM.

Bathe K 1996 *Finite element procedure*. Prentice Hall.

Ben Dhia H 1998 Multiscale mechanical problems: the Arlequin method. *Comptes Rendus de l'Academie des Sciences Series IIB Mechanics Physics Astronomy*, **326**(12), 899–904.

Ben Dhia H 1999 Numerical modelling of multiscale problems: the Arlequin method. *CD Proceedings of ECCM'99, Munich*.

Ben Dhia H and Rateau G 2005 The Arlequin method as a flexible engineering tool. *International Journal for Numerical Methods in Engineering*, **62**(11), 1442–1462.

Biscani F, Giunta G, Belouettar S, and Carrera E 2011 Variable kinematic beam elements coupled via Arlequin method. *Composite Structures*, **93**(2), 697–708.

Blanco PJ, Feijoo RA, and Urquiza SA 2008 A variational approach for coupling kinematically incompatible structural models. *Computer Methods in Applied Mechanics and Engineering*, **197**, 1577–1620.

Brezzi F and Marini LD 2005 The three-field formulation for elasticity problems. *GAMM Mitteilungen*, **28**(1), 124–153.

Carrera E and Giunta G 2010 Refined beam theories based on a unified formulation. *International Journal of Applied Mechanics*, **2**(1), 117–143.

Fish J 1992 The s-version of the finite element method. *Computers & Structures*, **43**(3), 539–547.

Fish J and Markolefas S 1993 Adaptive s-method for linear elastostatics. *Computational Methods in Applied Mechanics and Engineering*, **103**, 363–396.

Fish J, Pan L, Belsky V, and Gomaa S 1996 Unstructured multigrid method for shells. *International Journal for Numerical Methods in Engineering*, **39**(7), 1181–1197.

Hu H, Belouettar S, Potier-Ferry M, and Daya EM 2008 Multi-scale modelling of sandwich structures using the Arlequin method Part I: Linear modelling. *Finite Elements in Analysis and Design*, **45**(1), 37–51.

Hu H, Belouettar S, Potier-Ferry M, Daya EM, and Makradi A 2010 Multi-scale nonlinear modelling of sandwich structures using the Arlequin method. *Composite Structures*, **92**(2), 515–522.

Möes N, Dolbow J, and Belytschko T 1999 A finite element method for crack growth without remeshing. *International Journal for Numerical Methods in Engineering*, **46**(1), 131–150.

Park KC and Felippa CA 2000 A variational principle for the formulation of partitioned structural systems. *International Journal for Numerical Methods in Engineering*, **47**(1–3), 395–418.

Reddy JN 2004 *Mechanics of Laminated Composite Plates and Shells*. CRC Press.

Szabó BA and Babuška I 1992 *Finite element analysis*. John Wiley & Sons, Inc.

12

Guidelines and recommendations

The beam models introduced so far are generally composed of full expansions to a certain order; this means that all the displacement variables of a given N-order theory were taken into account. Depending on the structural problem that has to be analyzed, the contribution of each term of a theory to the final solution varies; that is, some variables are more important than others in detecting the mechanical behavior of a structural system. Moreover, some terms could have no effectiveness at all, as their absence will not corrupt the accuracy of the solution. The present unified formulation permits us to investigate the role of each displacement variable of a given beam theory and to recognize which terms are effective and which are not. This chapter describes the so-called mixed axiomatic–asymptotic approach and gives general guidelines to determine the most adequate model for a given problem. Some examples are presented in details to evaluate reduced models and to highlight the role of characteristic parameters such as the slenderness ratio, the thickness, and the loading conditions.

12.1 Axiomatic and asymptotic methods

The development of a beam theory can be seen as a procedure to reduce a 3D problem to a 1D one. A displacement, stress, or strain variable, f, is described by means of one or more M additional variables f_τ ($\tau = 1, M$) defined at an assigned point on the beam section that, most of the time, coincides with the beam axis.

Beam Structures: Classical and Advanced Theories, First Edition. Erasmo Carrera, Gaetano Giunta and Marco Petrolo.

Base functions, $F_\tau(x, z)$, of the cross-section coordinates are exploited to express the variable expansions as in the following equation:

$$f(x, y, z) = F_\tau(x, z) f_\tau(y), \quad \tau = 1, 2, \ldots, M \tag{12.1}$$

where x and z are the cross-section coordinates, y is the beam-axis coordinate, and M indicates the number of terms in the expansion. A linear theory can be therefore obtained by adopting a three-component polynomial base:

$$\begin{aligned} u_x &= u_{x_1} + x\, u_{x_2} + z\, u_{x_3} \\ u_y &= u_{y_1} + x\, u_{y_2} + z\, u_{y_3} \\ u_z &= u_{z_1} + x\, u_{z_2} + z\, u_{z_3} \end{aligned} \tag{12.2}$$

where u_x, u_y, and u_z are the displacement components defined using nine variables $(u_{x_1}, u_{x_2}, \ldots, u_{z_3})$.

There are two main techniques that are commonly exploited to construct a structural model:

- the axiomatic hypothesis method;
- the asymptotic expansion method.

In the so called *axiomatic* method, the expression in Equation (12.2), derives from scientific intuition. Well-known examples of axiomatic-built models are those by Euler and Bernoulli (Euler, 1744) and Timoshenko (1921, 1922). The excellent review by Kapania and Raciti (1989) presents a comprehensive description of refined beam theories developed axiomatically.

The so-called *asymptotic* method exploits a perturbation parameter such as the span-to-height ratio to analyze the influence of each variable and then build a suitable model for a given structural problem. The 3D problem is reduced to a 1D model by means of an asymptotic series of the characteristic parameter and retaining those terms that exhibit the same order of magnitude as the perturbation parameter when it vanishes. Two of the most important pioneers of the asymptotic technique were Cicala (1965) and Gol'denweizer (1962), whereas a reference asymptotic model for beams is the one developed by Yu and Hodges (2005).

12.2 The mixed axiomatic–asymptotic method

The advantages and drawbacks due to the axiomatic and asymptotic approaches have been largely discussed in the open literature. While the former can construct models having a large number of displacement variables, the latter provides solutions which are usually strongly problem dependent; this aspect implies the need for different reduced models for different problems. Moreover, the axiomatic method is relatively simpler to use but is characterized by a certain lack of information about the convergence of increasing order terms to the 3D solution; the asymptotic one is systematic but more cumbersome, especially when several parameters have to be taken into account (thickness, boundary and loading conditions, slenderness ratio, etc.). The present beam formulation could be seen as an alternative "third" solution to develop refined beam models by simply fixing the order of the expansion for the cross-section coordinates, N, with no restrictions, as already suggested by Washizu (1968). Such a solution was not considered in the past because it implies an arbitrary increase in the number of governing equations to be solved.

A further analysis tool offered by the present beam formulation is related to the implementation of the so-called mixed axiomatic–asymptotic method that was proposed by Carrera and Petrolo (2011). This method permits us to obtain asymptotic-like results starting from axiomatic-like hypotheses. This procedure is briefly described below:

1. The problem data are fixed $\Rightarrow$ *Loadings, boundary conditions materials*
2. A set of output variables is chosen $\Rightarrow$ σ, ϵ, u
3. The CUF is used to generate the governing equations for the considered theories $\Rightarrow$ $u = F_\tau \, u_\tau$
4. A theory is fixed and used to establish the accuracy

$\Rightarrow$

$N=0$	$N=1$		$N=2$		
u_{x_1}	$u_{x_2}x$	$u_{x_3}z$	$u_{x_4}x^2$	$u_{x_5}xz$	$u_{x_6}z^2$
u_{y_1}	$u_{y_2}x$	$u_{y_3}z$	$u_{y_4}x^2$	$u_{y_5}xz$	$u_{y_6}z^2$
u_{z_1}	$u_{z_2}x$	$u_{z_3}z$	$u_{z_4}x^2$	$u_{z_5}xz$	$u_{z_6}z^2$

5. Each term is deactivated in turn

$\Rightarrow$

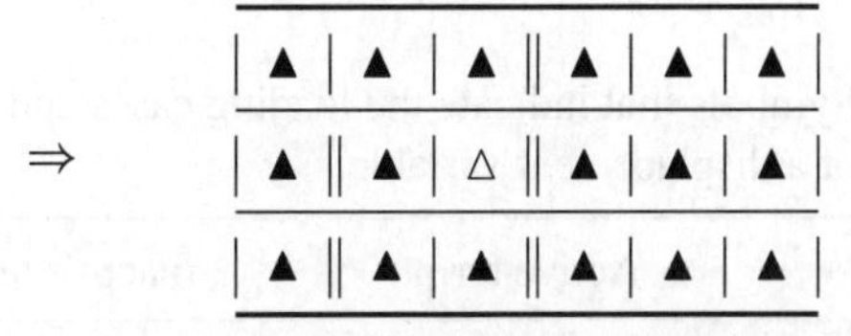

▲	▲	▲	▲	▲	▲
▲	▲	△	▲	▲	▲
▲	▲	▲	▲	▲	▲

6. Does the absence of the term corrupt the solution?

$\Rightarrow$

Active term	Inactive term
Yes, ▲	No, △

Table 12.1 Locations of the displacement variables within the table layout.

$N=0$	$N=1$		$N=2$		
u_{x_1}	$u_{x_2}x$	$u_{x_3}z$	$u_{x_4}x^2$	$u_{x_5}xz$	$u_{x_6}z^2$
u_{y_1}	$u_{y_2}x$	$u_{y_3}z$	$u_{y_4}x^2$	$u_{y_5}xz$	$u_{y_6}z^2$
u_{z_1}	$u_{z_2}x$	$u_{z_3}z$	$u_{z_4}x^2$	$u_{z_5}xz$	$u_{z_6}z^2$

This table shows the layout notation that defines the position of each displacement variable inside the table in the case of a second-order model, $N = 2$.

7. The most suitable kinematics model is then detected for a given structural lay-out

△	▲	△	▲	△	▲
▲	▲	△	△	▲	△
△	▲	▲	△	▲	△

$$\Rightarrow \quad \begin{aligned} u_x &= x\,u_{x_2} + x^2\,u_{x_4} + z^2\,u_{x_6} \\ u_y &= u_{y_1} + x\,u_{y_2} + xz\,u_{y_5} \\ u_z &= x\,u_{z_2} + z\,u_{z_3} + xz\,u_{z_5} \end{aligned}$$

A graphic notation is introduced to improve the readability of the results. Table 12.1 shows the locations held by each second-order beam model term within the tabular layout. The first column presents the constant terms, $N = 0$, the second and third columns the linear terms, $N = 1$, and the last three columns show the parabolic terms, $N = 2$. Each term can be activated or deactivated as shown in Table 12.2 where the loading case symbols are also shown. According to the

Table 12.2 Symbols that indicate the loading cases and the presence of a displacement variable.

Loading case	Active term	Inactive term
Bending	▲	△
Torsion	●	○
Axial	▼	▽

This table shows the symbols that are used to indicate the presence of a displacement variable and the loading case considered.

Table 12.3 Symbolic representation of the reduced kinematic model with u_{y_3} deactivated.

▲	▲	▲	▲	▲	▲
▲	▲	△	▲	▲	▲
▲	▲	▲	▲	▲	▲

This table shows the graphic representation of a second-order beam model having u_{y_3} deactivated.

notation adopted, the beam model given by Table 12.3 is used for the bending analysis and it refers to the following cross-section displacement field:

$$\begin{aligned} u_x &= u_{x_1} + x\, u_{x_2} + z\, u_{x_3} + x^2\, u_{x_4} + xz\, u_{x_5} + z^2\, u_{x_6} \\ u_y &= u_{y_1} + x\, u_{y_2} + \qquad\quad + x^2\, u_{y_4} + xz\, u_{y_5} + z^2\, u_{y_6} \\ u_z &= u_{z_1} + x\, u_{z_2} + z\, u_{z_3} + x^2\, u_{z_4} + xz\, u_{z_5} + z^2\, u_{z_6} \end{aligned} \tag{12.3}$$

Example 12.2.1 *Let us evaluate the influence of the u_{z_3} term of a fourth-order beam model in a torsional analysis of the thin-walled cylinder shown in Figure 8.5 with L/d equal to* 10. *The torsion is investigated through the application of two concentrated forces, F_y, at* [$\pm\, d/2$, 0, L]. *The magnitudes of the forces are* $\mp$250 [kN]. *Table 12.4 shows the reduced beam model considered. The influence of the term is evaluated by means of the percentage variations*

$$\delta_u = \frac{u}{u_{N=4}} \times 100,\ \delta_\sigma = \frac{\sigma}{\sigma_{N=4}} \times 100 \tag{12.4}$$

that are computed with respect to the full fourth-order, $N = 4$, beam model. Table 12.5 shows the results referred to the three displacement variables computed at [$d/2$, 0, L]. *It can be seen how u_{z_3} influences u_x and u_z yet it does not affect u_y.*

Table 12.4 Reduced fourth-order model for the torsional analysis of a thin-walled cylinder.

•	•	•	•	•	•	•	•	•	•	•	•	•	•	•
•	•	•	•	•	•	•	•	•	•	•	•	•	•	•
○	•	•	•	•	•	•	•	•	•	•	•	•	•	•

This table presents a reduced fourth-order beam model where the u_{z_3} variable is deactivated.

Table 12.5 Effect of the absence of u_{z_3} on different output variables.

δ_{u_x} (%)	δ_{u_y} (%)	δ_{u_z} (%)
96.9	100.0	118.8

This table presents the effects caused by the absence of u_{z_3} on different output variables.

Example 12.2.1 illustrates the procedure for investigating the effect of a single term on different output variables. This procedure is used to determine all the ineffective terms of a refined model for a given structural problem.

12.3 Load effect

Bending, torsion, and traction load cases are considered for a compact square cantilevered beam. An $N = 4$ beam model is considered as the reference solution to evaluate the effectiveness of each displacement variable in detecting the displacement components. Table 12.6 shows the set of terms that is needed to detect the fourth-order accuracy: 11 out of 45 terms are needed and the explicit expression of the beam model is

$$\begin{aligned} u_x &= xz\, u_{x_5} + x^3 z\, u_{x_{12}} + xz^3\, u_{x_{14}} \\ u_y &= z\, u_{y_3} + x^2 z\, u_{y_8} + z^3\, u_{y_{10}} \\ u_z &= u_{z_1} + z^2\, u_{z_6} + x^4\, u_{z_{11}} + x^2 z^2\, u_{z_{13}} + z^4\, u_{z_{15}} \end{aligned} \tag{12.5}$$

the torsion-related reduced model is presented in Table 12.7; in this case 9 out of 45 terms are needed and the beam model is given by

$$\begin{aligned} u_x &= z\, u_{x_3} + x^2 z\, u_{x_8} + z^3\, u_{x_{10}} \\ u_y &= xz\, u_{y_5} + x^3 z\, u_{y_{11}} + xz^3\, u_{y_{14}} \\ u_z &= x\, u_{z_2} + x^3\, u_{z_7} + xz^2\, u_{z_9} \end{aligned} \tag{12.6}$$

Table 12.6 Set of active displacement variables for the bending analysis of a square cross-section beam.

$M_{eff}/M = 11$														
△	△	△	△	▲	△	△	△	△	△	△	▲	△	▲	△
△	△	▲	△	△	△	△	▲	△	▲	△	△	△	△	△
▲	△	△	△	△	▲	△	△	△	△	▲	△	▲	△	▲

This table presents the set of active terms to determine the bending behavior of a square cross-section beam; 11 out of 45 terms are needed.

Table 12.7 Set of active displacement variables for the torsional analysis of a square cross-section beam.

$M_{eff}/M = 9$														
○	○	●	○	○	○	○	●	○	●	○	○	○	○	○
○	○	○	○	●	○	○	○	○	○	○	●	○	●	○
○	●	○	○	○	○	●	○	●	○	○	○	○	○	○

This table presents the set of active terms to determine the torsional behavior of a square cross-section beam; 9 out of 45 terms are needed.

Finally, the traction case is shown in Table 12.8:

$$
\begin{aligned}
u_x &= x\, u_{x_2} + x^3\, u_{x_7} + xz^2\, u_{x_9} \\
u_y &= u_{y_1} + x^2\, u_{y_4} + z^2\, u_{y_6} + x^4\, u_{y_{11}} + x^2z^2\, u_{y_{13}} + z^4\, u_{y_{15}} \\
u_y &= z\, u_{z_3} + x^2 z\, u_{z_8} + z^3\, u_{z_{10}}
\end{aligned}
\tag{12.7}
$$

It is important to emphasize how the beam models in Equations (12.5, 12.6, and 12.7) are substantially different to each other this means that each loading case needs its own reduced beam model. The combined reduced beam model necessary to detect the fourth-order solution for bending, torsion, and traction loads is presented in Table 12.9.

12.4 Cross-section effect

Another important parameter for determining a refined beam model is represented by the geometry of the cross-section. General guidelines say that compact beams need less cumbersome models than thin-walled ones. Other aspects are the

Table 12.8 Set of active displacement variables for the axial analysis of a square cross-section beam.

$M_{eff}/M = 12$														
▽	▼	▽	▽	▽	▽	▼	▽	▼	▽	▽	▽	▽	▽	▽
▼	▽	▽	▼	▽	▼	▽	▽	▽	▽	▼	▽	▼	▽	▼
▽	▽	▼	▽	▽	▽	▽	▼	▽	▼	▽	▽	▽	▽	▽

This table presents the set of active terms to determine the axial behavior of a square cross-section beam; 12 out of 45 terms are needed.

Table 12.9 Combined set of active displacement variables for the bending, torsional, and traction analysis of a square cross-section beam.

$M_{eff}/M = 32$														
	▼	•		▲		▼	•	▼	•		▲		▲	
▼		▲	▼	•	▼		▲•		▲	▼	•	▼	•	▼
▲	•	▼			▲	•	▼	•	▼	▲		▲		▲

This table presents the combined set of active terms to determine the bending, torsional, and traction behavior of a square cross-section beam; 32 out of 45 terms are needed.

symmetry/asymmetry and the presence of closed/open sections. Two cross-section geometries are analyzed here: annular and airfoil shaped. In both cases a torsional load is applied to the free tip. Table 12.10 shows the reduced beam models equivalent to a full fourth-order model that is expressed by

$$\begin{aligned} u_x &= x\, u_{x_2} + z\, u_{x_3} + x^3\, u_{x_7} + x^2 z\, u_{x_8} + x z^2\, u_{x_9} + z^3\, u_{x_{10}} \\ u_z &= u_{y_1} + x^2\, u_{y_4} + xz\, u_{y_5} + z^2\, u_{y_6} + x^4\, u_{y_{11}} + x^3 z\, u_{y_{12}} \\ &\quad + x^2 z^2\, u_{y_{13}} + x z^3\, u_{y_{14}} + z^4\, u_{y_{15}} \\ u_y &= x\, u_{z_2} + z\, u_{z_3} + x^3\, u_{z_7} + x^2 z\, u_{z_8} + x z^2\, u_{z_9} + z^3\, u_{z_{10}} \end{aligned} \tag{12.8}$$

Table 12.11 shows the equivalent result for the airfoil-shaped cantilevered beam. In this case all the 45 displacement variables are needed, that is, each term of the fourth-order expansion plays a role in detecting the mechanical behavior of the structure.

Table 12.10 Set of active displacement variables for the torsional analysis of a thin-walled annular beam.

$M_{eff}/M = 21$														
▽	▼	▼	▽	▽	▽	▼	▼	▼	▼	▽	▽	▽	▽	▽
▼	▽	▽	▼	▼	▼	▽	▽	▽	▽	▼	▼	▼	▼	▼
▽	▼	▼	▽	▽	▽	▼	▼	▼	▼	▽	▽	▽	▽	▽

This table presents the set of active terms to determine the torsional behavior of a thin-walled annular beam; 21 out of 45 terms are needed.

Table 12.11 Set of active displacement variables for the torsional analysis of an airfoil-shaped beam.

$M_{eff}/M = 45$														
•	•	•	•	•	•	•	•	•	•	•	•	•	•	•
•	•	•	•	•	•	•	•	•	•	•	•	•	•	•
•	•	•	•	•	•	•	•	•	•	•	•	•	•	•

This table presents the set of active terms to determine the torsional behavior of an airfoil-shaped beam; 45 out of 45 terms are needed.

12.5 Output location effect

The reduced models seen so far were built by considering a certain output variable at a given point. In this section, a given output variable is computed at different spanwise locations and the respective reduced models are then evaluated. Two spanwise locations are considered, $L/4$ and L. The results are shown in Table 12.12. It can been observed how the spanwise location plays an important role in determining the reduced model. In particular, the proximity of the boundary conditions increases the total number of variables needed.

Table 12.12 Set of active displacement variables for a thin-walled annular beam under torsion at different spanwise locations.

$x = 0, y = -h/2, z = L, M_{eff}/M = 21$														
○	•	•	○	○	○	•	•	•	•	○	○	○	○	○
•	○	○	•	•	•	○	○	○	○	•	•	•	•	•
○	•	•	○	○	○	•	•	•	•	○	○	○	○	○
$x = 0, y = -h/2, z = L/4, M_{eff}/M = 33$														
○	•	•	○	•	○	•	•	•	•	○	•	○	•	○
•	○	•	•	•	•	○	•	○	•	•	•	•	•	•
•	•	•	•	○	•	•	•	•	•	•	○	•	○	•

This table presents the set of active terms to determine the torsional behavior of a thin-walled annular beamat different spanwise locations.

Table 12.13 Set of active displacement variables to accomplish a given accuracy requirement for an airfoil-shaped beam under torsion.

$\overline{\delta_u} = 0\%, M_{eff}/M = 45$														
•	•	•	•	•	•	•	•	•	•	•	•	•	•	•
•	•	•	•	•	•	•	•	•	•	•	•	•	•	•
•	•	•	•	•	•	•	•	•	•	•	•	•	•	•
$\overline{\delta_u} \leq 15\%, M_{eff}/M = 42$														
•	•	•	•	•	•	•	•	•	•	•	•	○	○	•
•	•	•	•	○	•	•	•	•	•	•	•	•	•	•
•	•	•	•	•	•	•	•	•	•	•	•	•	•	•
$\overline{\delta_u} \leq 35\%, M_{eff}/M = 25$														
•	•	•	○	•	○	○	•	○	•	○	•	○	○	○
•	•	○	•	○	•	•	○	○	○	•	○	○	○	○
•	•	•	•	•	○	•	•	•	•	•	•	○	•	○

This table presents the set of active terms to determine the torsional behavior of an airfoil-shaped beam with a given accuracy.

12.6 Reduced models for different error inputs

Reduced models able to exactly detect a fourth-order solution have been considered so far. Another important option offered by the present beam formulation is represented by the possibility of choosing the accuracy range of the refined model and then determining the set of terms needed to accomplish the accuracy input. The analysis is conducted on the airfoil-shaped cantilevered beam under torsional and bending loads. The input parameter is the error with respect to the fourth-order solution:

$$\overline{\delta_u} = \left\| \frac{u - u_{N=4}}{u_{N=4}} \right\| \times 100 \tag{12.9}$$

For instance, $\overline{\delta_u} = 0$ implies that the exact $N = 4$ solution is sought. Table 12.13 shows the torsion-related results, while in Table 12.14 the bending case is addressed. In both cases, significant reductions in the total number of variables are

Table 12.14 Set of active displacement variables to accomplish a given accuracy requirement for an airfoil-shaped beam under bending.

$\overline{\delta_u} = 0\%, M_{eff}/M = 45$														
▲	▲	▲	▲	▲	▲	▲	▲	▲	▲	▲	▲	▲	▲	▲
▲	▲	▲	▲	▲	▲	▲	▲	▲	▲	▲	▲	▲	▲	▲
▲	▲	▲	▲	▲	▲	▲	▲	▲	▲	▲	▲	▲	▲	▲
$\overline{\delta_u} \leq 15\%, M_{eff}/M = 23$														
▲	▲	▲	▲	▲	△	△	▲	△	△	▲	▲	△	△	△
▲	▲	△	▲	▲	△	△	△	△	△	▲	△	△	△	△
▲	▲	▲	▲	▲	△	▲	△	▲	▲	△	△	▲	▲	△
$\overline{\delta_u} \leq 35\%, M_{eff}/M = 9$														
▲	△	▲	△	▲	△	△	△	△	△	△	△	△	△	△
▲	▲	△	△	△	△	△	△	△	△	△	△	△	△	△
△	▲	▲	△	▲	△	△	△	△	▲	△	△	△	△	△

This table presents the set of active terms to determine the bending behavior of an airfoil-shaped beam with a given accuracy.

observed with a totally different reduced model required for the torsion and the bending loading case. This represents further confirmation that the development of reduced higher-order beam models is strongly problem dependent. The possibility of dealing with full arbitrary-order models that is given by the present unified formulation is thus fundamental for analyzing structures of engineering interest where different loads, geometries, and boundary conditions are usually present simultaneously. It has to be emphasized that the present analysis considered isotropic materials only. In the case of composite materials, other important parameters such as the orthotropic ratio and the stacking sequence are expected to have the same role in determining the reduced models of those presented above. This aspect was shown by Carrera and Petrolo (2010) in the framework of refined plate models.

References

Carrera E and Petrolo M 2010 Guidelines and recommendations to construct theories for metallic and composite plates. *AIAA Journal*, **48**(12), 2852–2866.

Carrera E and Petrolo M 2011 On the effectiveness of higher-order terms in refined beam theories. *Journal of Applied Mechanics*, **78**(2), DOI: 10.1115/1.4002207.

Cicala P 1965 *Systematic approximation approach to linear shell theory*, Levrotto e Bella.

Euler L 1744 De curvis elasticis. *In: Methodus Inveniendi Lineas Curvas Maximi Minimive Proprietate Gaudentes, Sive Solutio Problematis Isoperimetrici Lattissimo Sensu Accepti*, Bousquet.

Gol'denweizer AL 1962 Derivation of an approximate theory of bending of a plate by the method of asymptotic integration of the equations of the theory of elasticity. *Prikladnaya Matematika i Mekhanika* **26**, 1000–1025.

Kapania K and Raciti S 1989 Recent advances in analysis of laminated beams and plates, part I: Shear effects and buckling. *AIAA Journal*, **27**(7), 923–935.

Timoshenko SP 1921 On the corrections for shear of the differential equation for transverse vibrations of prismatic bars. *Philosophical Magazine*, **41**, 744–746.

Timoshenko SP 1922 On the transverse vibrations of bars of uniform cross section. *Philosophical Magazine*, **43**, 125–131.

Washizu, W 1968 *Variational methods in elasticity and plasticity*. Pergamon.

Yu W and Hodges DH 2005. Generalized Timoshenko theory of the variational asymptotic beam sectional analysis. *Journal of the American Helicopter Society*, **50**(1), 46–55.

Index

Beam Structures: Classical and Advanced Theories, First Edition. Erasmo Carrera, Gaetano Giunta and Marco Petrolo.